Congenital Anomalies in Newborn Infants - Clinical and Etiopathological Perspectives

Edited by Rita P. Verma

Published in London, United Kingdom

IntechOpen

Supporting open minds since 2005

Congenital Anomalies in Newborn Infants - Clinical and Etiopathological Perspectives
http://dx.doi.org/10.5772/intechopen.83128
Edited by Rita P. Verma

Contributors
Rameshwar Prasad, Labaran Dayyabu Aliyu, Mehmet Semih Demirtaş, Jessie R. Maxwell, Walter Dehority, Hellen Ko, Richa Verma, Mohammad Hossein Khosravi, Bita Najafian, David Chitayat, Pierre Sinajon, Riyana Hirji, Saeed Sepehrnia, Rita P. Verma

First published in London, United Kingdom, 2021 by IntechOpen
IntechOpen is the global imprint of INTECHOPEN LIMITED, registered in England and Wales, registration number: 11086078, 5 Princes Gate Court, London, SW7 2QJ, United Kingdom
Printed in Croatia

British Library Cataloguing-in-Publication Data
A catalogue record for this book is available from the British Library

Additional hard and PDF copies can be obtained from orders@intechopen.com

Congenital Anomalies in Newborn Infants - Clinical and Etiopathological Perspectives
Edited by Rita P. Verma
p. cm.
Print ISBN 978-1-83962-988-4
Online ISBN 978-1-83962-989-1
eBook (PDF) ISBN 978-1-83962-990-7

For EU product safety concerns:
IN TECH d.o.o., Prolaz Marije Krucifikse Kozulić 3, 51000 Rijeka, Croatia, info@intechopen.com or visit our website at intechopen.com.

Meet the editor

Professor Rita Prasad Verma is the director of Neonatal-Perinatal Medicine and research at the Nassau University Medical Centers in New York, USA. She completed her medical education in India and trained in pediatrics and neonatal-perinatal medicine in the United States. She has performed animal and human research in neonatal medicine, particularly on fluid and electrolyte metabolism in prematurity, and published the results in more than 120 articles and abstracts in leading journals worldwide. She has delivered numerous invited keynote lectures nationally and internationally and has served as chief guest at international meetings. She is one of the pioneers in research on early life physiological weight loss in extremely and moderately preterm infants. She serves as a reviewer for major national and international journals and is a member of the editorial board of several journals. She is an active member of the American Academy of Pediatrics and its subcommittees and contributes to their policy and procedure reviews.

Contents

Preface

Congenital anomalies constitute a large group of diverse biochemical, functional, and structural defects presenting at birth. They are caused by a myriad of inherently unrelated etiopathogenic factors, categorized into genetic, environmental, infectious, nutritional, and metabolic. In a significant number of cases, the etiology remains unidentified. The malformations may be isolated, involve a single organ system, or be multisystemic. The defect may be benign, not requiring any intervention, or severe enough to be fatal at or after birth, the life span depending variably upon the diagnosis. The outcome may range from inconsequential to devastating and impart immense medical, social, emotional, and financial burden to the family. The management might be medical, surgical, none, or both, and the surgical procedure lifesaving or mere cosmetic. With such a broad spectrum of clinical presentation, management principles, and socio-economic implications, including all deformities in a single volume is beyond the scope of this book.

We begin with an introductory chapter that includes epidemiology, clinical evaluation, investigational procedures, and the recurrence risk assessment of congenital anomalies. The following two chapters discuss the etiopathogenesis of congenital malformations and highlight the diversity in their origin. The successive chapters review some of the most complex malformations selected from the major organ systems for their anatomy, pathophysiology, and current management principles. Finally, to underscore the geographical variabilities, we include a chapter on the demographic analysis of the common congenital malformations in Africa. I am grateful to the esteemed authors for their time and effort in sharing their expertise on the selected topics. Without their collaboration, this book would not have been possible. I am also grateful to the entire team of publishers that have tirelessly worked towards the book's successful publication.

I dedicate this effort to my father, Mr. Shib Chandra Prasad, and my mother, Mrs. Sita Pati Prasad, since deceased, for their support, role modeling, and unconditional love. Our journey together was brief but most formative and impactful.

Rita Prasad Verma, M.D., F.A.A.P.
Professor of Clinical Pediatrics,
Nassau University Medical Center,
East Meadow, NY, USA

Chapter 1

Introductory Chapter: Epidemiology, Evaluation and Risk Assessment of Congenital Anomalies

Rita Prasad Verma

1. Introduction

Congenital anomalies (CA) are the leading cause of infant mortality in the USA (**Figure 1**) [1]. The five most common birth malformations recorded in the USA are club foot (1 in 593 births), Down syndrome (1 in 707 births), pulmonic stenosis/atresia (1 in 1052 births), cleft palate (1 in 1687 births), and limb defects (1 in1943 birth). Congenital heart disorders, neural tube defects, and Down syndrome are the three commonest causes of mortality due to CA in that order

One in every 33 children born in the USA suffers from a birth anomaly, and 1 in 5 deaths among infants is due to morbidities related to them [2]. Globally about 7.9 million children are born with a major birth malformation every year, and CA is the fourth leading cause of neonatal death worldwide [3]. The annual cost of hospitalization due to congenital malformations is reported to be $22.9 billion in the USA [4]. The cost of care of infants suffering from CA is relatively higher than any

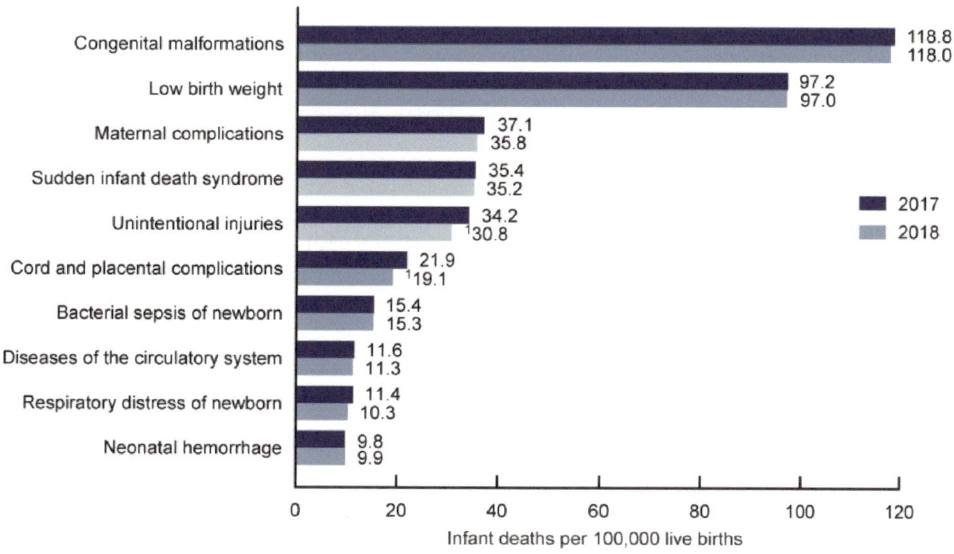

Figure 1.
Infant mortality per 100,000 live births in the United States.

other childhood morbidity. While 3.0% of all hospitalizations are attributed to birth anomalies, the diagnosis accounts for 5.2% of total hospital costs. The cost of care is highest in the children who are <1 year of age and accounts for 35.0% of the total hospitalization costs among the pediatric population. The congenital cardiac defects register the highest CA-associated hospitalization costs across all age groups at 26.6%, while only 14.0% of CA–associated hospitalizations are attributed to this system.

2. Abnormal morphogenesis

Normal morphogenesis is a systematic process that involves a combination of several simultaneously or sequentially occurring histo-physiological activities at the cellular and molecular levels. These processes are cell migration, aggregation of identical cell types, the interaction between neighboring tissues, controlled cell death, mechanical forces, and hormonal effects. The congenital anomaly is an outcome of abnormal morphogenesis and is defined as any structural or functional anomaly detected at birth that interferes with the performance or appearance of the subject. The etiopathogenesis of congenital anomalies is broadly divided into genetic and non-genetic groups, and the latter one is classified into several major sub-categories. Some anomalies remain idiopathic in etiopathogenesis, and others may be multifactorial, i.e., have both genetic and environmental influences. Congenital anomalies are categorized as major and minor as per their clinical implications. While major anomalies have significant adverse medical consequences, minor malformations may be of cosmetic significance only. The prevalence of major malformations is reported to range from 2 to 4 percent according to the population studied, while minor anomalies may be as high as 35% in certain populations [5]. The presence of 3 or more minor malformations is associated with an increased risk of a major anomaly or syndrome. Major anomalies are generally the consequence of molecular defects that interfere with normal morphogenesis processes like apoptosis, intracellular signaling, migration of neural crest derivatives, and chromatin remodeling. Some of the genes (e,g. Homeobox genes in Synpolydactyly, microphthalmia, and holoprosencephaly), transcription factors (e.g., deletions of T-box 1 in Conotruncal heart defects of DiGeorge syndrome), fibroblast growth factor receptors (in Craniosynostoses syndromes), and enzyme defects (such as cholesterol biosynthesis leading to Smith-Lemli-Opitz syndrome) have been identified as etiopathogenic factors for specific major malformations.

2.1 Patterns of abnormal phenotypes

Abnormal morphogenesis may manifest in various forms at the macroscopic and microscopic levels [6]. Following are some of the common patterns and terminologies used in the description of an abnormal phenotype.

- Anomaly is defined as a significant morphological or anatomic variation in the phenotype from the standard reference population. The difference may be macroscopic or microscopic. Anomalies are categorized into major and minor. Major anomalies have significant adverse implications in functional, anatomical, psychological, social, or cosmetic wellbeing, and minor anomalies are variants with no significant medical or major cosmetic consequences. The presence of three or more minor anomalies suggests significant defects in morphogenesis. Even though detected in only 0.5% of births, almost 90% of such cases have ≥1 major anomaly and might eventually end up with the diagnosis of some associations or syndrome.

- Malformation is an inherently non-progressive morphologic anomaly brought about primarily by an intrinsic error in the developmental process at the cellular or molecular levels of an organ or a body part. Malformations may be isolated, or part of a syndrome and are causally heterogeneous. The aberration can happen due to gene mutations, exposure to teratogens, or a combination of the two. For example, the limb anomaly in Holt–Oram syndrome is genetic in origin, whereas in thalidomide embryopathy, it is due to teratogenic exposure.

- Deformation alludes to a distortion of the shape or size in an otherwise normal body part of a normally developing or developed fetus, caused by some aberrant extrinsic or intrinsic mechanical forces. Examples include craniofacial asymmetry, arthrogryposis, and metatarsus adductus (**Figure 2**). Uterine anomalies and abnormal fetal positions are important setups for the generation of fetal deformation. Deformations are causally heterogeneous and may result in the loss of alignment or abnormal positioning and distorted configuration. Deformations are generally reversible after birth but may be difficult if the transduction of abnormal mechanical forces is prolonged. Deformation generally happens after organogenesis, but if in early gestation, it may permanently alter the structural relationships. If mechanical transduction happens during embryogenesis, anomalies in the neural tube, tendons, and joints may be produced.

- Disruption is defined as a static morphologic abnormality brought about in-utero by some destructive mechanical forces acting upon an otherwise normally developing or developed fetal tissue or physical part. It results in the destruction of the involved body part and may cause the developmental arrest of the adjacent tissues, thereby leading to a secondary malformation (**Figure 3**). It can be an initial event in a sequence of events if it occurs early in gestation. It is causally heterogeneous and may be isolated or part of a syndrome or other broader patterns. A disruption can impart a particularly distinctive appearance because of the loss of tissue and aberrant differentiation of adjacent tissues with or without the production of adhesions. The process results in cell death and tissue destruction, and the mechanisms may include vascular compromise, anoxia, teratogens exposure, infections, or mechanical forces. Clinical entities such as missing digits or limbs are examples of disruption. The process

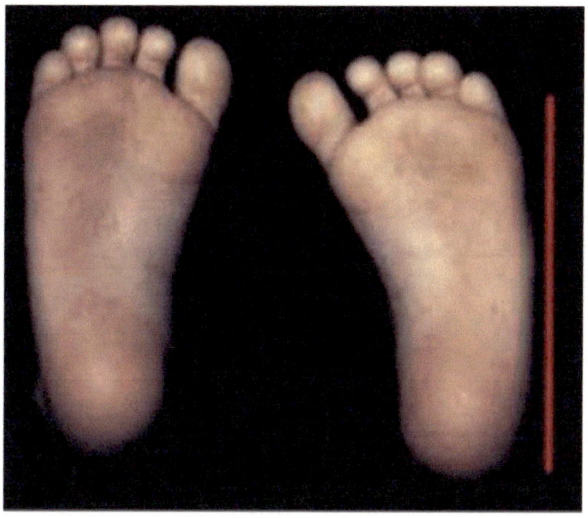

Figure 2.
Metatarsus adductus as an example of deformation due to intrauterine positioning or space constraints.

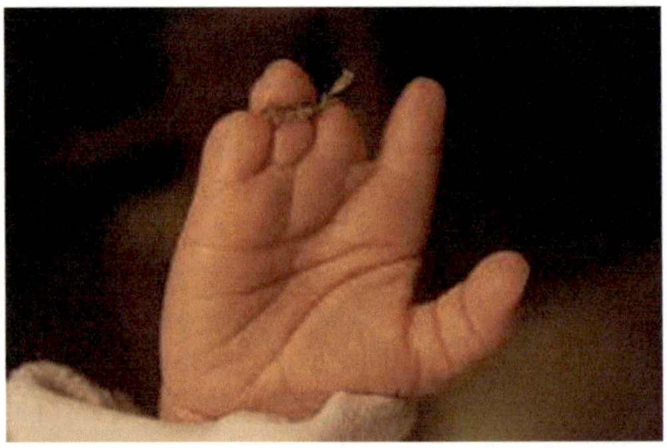

Figure 3.
Amputation of digits with a constricting amniotic band as an example of disruption.

characteristically affects several tissue types in a specific anatomical region, and the phenotypic abnormalities may cross cell lines of embryonic development. Some pathological developmental processes can cause both disruption and deformation. For example, constriction rings at the tip of a finger associated with bands (fibrous strands of tissue) are often used as an example of disruption, but fibrous bands can also cause deformation. "Amniotic bands" encircling a limb are one possible mechanism of disruption of an extremity.

- Dysplasia is an abnormality in the growth and development of cell or tissue histology, or the anatomical structure or physiological function resulting from such growth. Tumors and malignancies are the results of a dysplasia process. Dysplasia can be isolated in occurrence or be a part of broader patterns. Dysplasias are causally heterogeneous and can be triggered by genetic factors, teratogens exposure, or metabolic disorders. The presentation may involve one or multiple germ layers and single or multiple organs. It may be localized or generalized; unilateral or bilateral; focal or multifocal; benign, malignant or premalignant; static or progressive; or evanescent. Dysplasia may be associated with malformations. It can happen at the cellular level (microscopic), examples being BPD and fibrous epithelial dysplasia, or in organs (macroscopic) such as renal dysplasia. It may be a dynamic or an ongoing process and maybe widespread or confined to a single organ. It can occur both prenatally and postnatally.

- Sequence refers to the phenomenon of a single or multiple morphological anomalies cascading from a single primary anomaly, which might be malformation, disruption, dysplasia, or deformation in character. The resultant anomaly is not an essential and direct derivation of the primary cause, as happens in genetic aberration cases. A sequence can occur as an isolated phenomenon or as a component manifestation of broader patterns such as syndromes; and like malformations is causally heterogeneous. A common example is the Pierre Robin sequence, in which a small mandible, the primary anatomical defect, leads to protruding tongue, which in turn may interfere with the palatal closure and consequently create a cleft palate. In describing a sequence, sometimes it may be difficult to distinguish between the primary and the consequential effects.

- Association is defined as a pattern of morphological anomalies which are not causally related but occur together more often than would be expected by chance only. Some associations may be syndromes with overlapping features. Such cases may eventually be identified to have a pathological etiology and then moved to the category of a syndrome. In the case of CHARGE association, after the causative gene was identified, the nomenclature was changed to CHARGE syndrome.

- A syndrome is a combination of causally but not necessarily pathogenically related anomalies that are characterized into a specific condition. The anomalies can be malformations, deformations, disruptions, sequences or dysplasia, major or minor, or functional, such as those affecting the neurological, cognitive, sensory, or behavioral performances.

2.2 Anomalies of genotype

Genetically determined and inheritable diseases can occur as a result of chromosomal or single-gene (Mendelian) abnormalities. Some of the inherited disorders are multifactorial in origin, and others may be attributed to defects in the mitochondrial chromosomes [7]. The last category has the highest phenotypic variability due to heteroplasmy and the fact that mitochondrial DNA has high incidence of mutation [8]. As the mitochondrial DNA is inherited from the mother, the genetic transmission occurs from the affected mother only. Single gene disorders (SGD) may present with autosomal dominant, autosomal recessive, X-linked recessive, and X-linked dominant patterns of inheritance. Cystic fibrosis is the commonest SGD reported in the Caucasian population.

Chromosomal anomalies may result from maldistribution (numerical aberrations) or rearrangements (structural abnormalities) of chromosomes. A numerical error in the array of chromosomes is termed aneuploidy, which presents as polyploidy with an addition or monosomy with a reduction in the number of chromosomes (**Figure 4**). Most of these conditions result from the failure of chromosomes to disjoin during meiosis (non-disjunction). Some of the common examples of aneuploidy are the trisomy syndromes or Turner and Klinefelter syndromes.

Structural chromosome disorders result from the breakage and rearrangement of segments or parts of a chromosome (**Figure 5**).

- Deletion refers to the loss of a piece or section of chromosomal material. If too small to be visualized under a microscope, it is termed microdeletion. Deletions can be terminal if only one break is present at the end or interstitial if two pieces of chromosome material are lost from within the chromosome. There is only one copy of a particular chromosome segment instead of the usual two copies in deletion syndromes.

- Duplication presents with an extra copy of a segment of a chromosome. So there are three copies of a particular chromosome segment instead of the usual two.

- Mutation indicates a change in the DNA sequence that leads to a change in its function. A mutation may be silent (no overt clinical signs or symptoms as amino acids may be encoded on different codons); missense (the codon is changed by a new nucleotide); nonsense (new nucleotide changes the codon to a STOP codon so that the mRNA translation is stopped) or splice-site when mutation at splice site prevents removal of an intron.

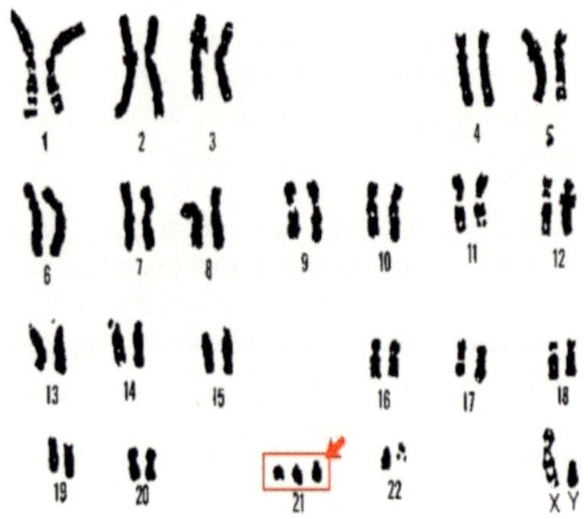

Figure 4.
An example of aneuploidy with Trisomy 21 showing an extra chromosome 21.

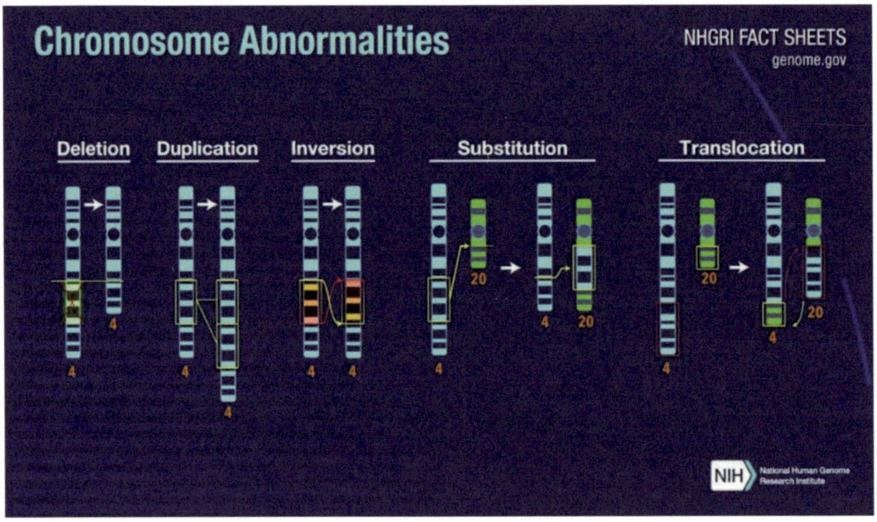

Figure 5.
Common structural chromosomal anomalies resulting in abnormal genotype.

- Translocation happens when genetic material is exchanged between two chromosomes. It may be balanced with no gain or loss of DNA and no anomalies or unbalanced when the process results in gain or loss of chromosomal material. Even though there may not be any phenotypical anomalies, balanced translocations might have implications for the offsprings of the patient.

- Reciprocal translocation is an anomaly in which the phenotype is normal despite the break and exchange of material between 2 chromosomes.

- Robertsonian translocation is another example in which the phenotype may be normal. It involves only selected chromosomes, such as 13,14,15,21,22. In this anomaly, the short arm is lost, and long arms fuse at the centromere.

- Inversion is described when a piece of a chromosome breaks at two places and then is reinserted in the opposite direction. Inversions may be paracentric if it does not involve centromere or pericentric if the centromere is involved.

- Isochromosomes result from abnormal mitosis in which a break at centromere results in two short (p) arms or two long (q) arms from the same side, both carrying identical genes.

- Dicentric chromosomes are defined as the abnormal fusion of two chromosome pieces, both having a centromere.

- Ring chromosomes form when the deletion happens at the ends of both arms of the same chromosome and the remaining chromosome join, making a ring-like shape. The chromosomes may be eventually lost, resulting in monosomy.

3. Evaluation of an infant with congenital malformations

The clinical evaluation of a child born with an anomaly begins with a detailed history of the extended family and a comprehensive history of the parents, with special attention to age, gravidity, parity, miscarriages, pregnancy-related complications, history of prescription drugs intake, substance abuse, significant illnesses and consanguinity among others. This is followed by a thorough physical examination, which should consist of a detailed assessment of craniofacial profile for dysmorphology and of individual organ systems, including the vertebral column, extremities, and skin. The histopathological features of the placenta and umbilical cord should be noted. The next step includes the performance of specific diagnostic tests, which are selected on the basis of the results of history and physical findings and should be individualized according to the case. Referral to a geneticist is indicated if one major or more than two minor anomalies are present.

3.1 Physical examination

The physical examination starts with the standard measurements of weight, length, and head circumference. In infants suffering from short stature, skeletal dysplasia, or suspected Marfan syndrome, arm span and lower segment/upper segment ratio are useful variables to note. The following is a list of specific features and characteristics to seek during the general evaluation of an infant with congenital malformations: midface hypoplasia; prognathism , retrognathia, micrognathia; facial asymmetry; hypertelorism, hypotelorism; ophthalmoplegia; esotropia, exotropia; cataract; nystagmus; ptosis; inner canthal distance, outer canthal distance, interpupillary distance; palpebral fissures length; long, anteriorly or posteriorly rotated ears; low set ears; microtia; prominent, bulbous nasal tip; slit appearance of nose; anteverted nares; long or smooth philtrum; macrostomia, microstomia, high arched palate, cleft lip/palate; cleft uvula,; macroglossia, protruding tongue ; wide or short neck , neck webbing ; pectus excavatum or pectus carinatum; wide-spaced nipples; scoliosis, deep sacral dimple, sacral hair tuft, sacral tag; sirenomelia; limited range of motion of extremities, contracture; polydactyly syndactyly, brachydactyly, arachnodactyly, broad thumbs and toes; clubfoot; breast development; Tanner staging. Ambiguous genitalia; micropenis; cryptorchidism; hypoplasia of labia or vaginal hypoplasia/atresia, sparse or excess body hair; abnormally light hair; skin hyperpigmentation /hypopigmentation; albinism; nail dystrophy. In addition to these, a detailed systemic evaluation is an essential and integral part of the physical examination of a child born with anomalies. The physical examination should be standardized and performed by a trained dysmorphologist [9].

3.2 Specific tests and laboratory investigations

As mentioned earlier, the need and selection of testing depend upon the results of the history and physical examination. Specific tests commonly include computed tomography (CT), and magnetic resonance imaging (MRI) scans of the central nervous system (CNS), echocardiogram, skeletal survey, and autopsy if the patient expires. An ultrasound evaluation of the genitourinary system is indicated in the conditions of ambiguous genitalia in order to detect the presence of abdominal gonads and uterus and to assess the anatomy of kidneys, ureters, bladder, and testicles. Fundoscopy may be done to detect ocular anomalies, such as retinal colobomas, optic disc defects, and chorioretinitis. TORCH infections and other significant viral infections should be ruled out by performing specific tests as clinically indicated.

3.3 Tests for suspected genetic disorders

The selection of chromosomal studies is guided by the clinical presentation. Exome sequencing helps in the identification of rare single-gene disorders and is indicated in the conditions of multiple, complex congenital anomalies with no otherwise identified genetic defect. Molecular-based chromosome microarray studies (array comparative genomic hybridization or aCGH) are utilized to diagnose microdeletions. aCGH is the first line of testing in children with multiple anomalies and intellectual disabilities [10]. It has now replaced Giemsa banding karyotype (G-banding) and fluorescent in situ hybridization (FISH) studies. aCGH testing may be indicated in the following conditions: one or more major anomalies, three or more minor anomalies, unexplained intellectual deficiency, dysmorphism with or without tone and intellectual deficits, unexplained failure to thrive, family history of birth defects, and multiple miscarriages [11]. The FISH test may be used to confirm a microdeletion or microduplication detected by aCGH and is generally used in the prenatal screening of cells in amniotic fluid. In the conditions of consanguinity or suspected uniparental disomy (UPD), where the chromosome material is coming from one parent only, single nucleotide polymorphisms (SNPs) is indicated. Another test, called Whole-exome sequencing (WES), is used in children with multiple structural anomalies associated with intellectual disability and/or seizures [12]. This technique uses next-generation sequencing and can simultaneously analyze the coding regions of almost 19,000 to 20,000 genes. While WES cannot be used for the detection of microdeletions or microduplications, whole-genome sequencing (WGS) can detect larger deletions or duplications, as well as triple repeat expansions; and mutations in deep intronic regions, regulatory regions that are outside of the coding regions, and untranslated gene regions.

Some genetic disorders are metabolic in origin and require metabolic studies in addition to the genetic workup. These studies include tests for blood amino acids and urine organic acids, specific tests for the peroxisomal disorder, including liver biopsy in certain conditions, assessment of serum cholesterol precursors, and lactic acid and pyruvic acid levels in the blood and cerebrospinal fluid, among others. Metabolomics uses mass spectrometry to detect and quantify small molecules in plasma for the diagnosis of inborn errors of metabolism. This procedure scientifically studies the small molecule substrates (<1.5 kDa) called metabolites which are products of the biochemical processes in cellular metabolism. One of the cellular functions that entails mRNA, gene expression, and proteomic analyses leads to the release of gene products in the cell. Metabolomics meets the challenge of systems biology and functional genomics by integrating genomics, transcriptomic and proteomic information. Thus metabolomic information provides a better understanding of cellular biology.

3.4 Clinical care and outcome

The clinical care of a child is done via a multi-disciplinary approach and involves a team of pediatrician, geneticist, neurologist, cardiologist, surgeon, orthopedics surgeon, dermatologist, infectious disease specialist, and other specialties as indicated. The involvement of the social worker is important for family and social support. The natural course, complications, and eventual clinical outcomes depend on the anomaly and may range from inconsequential to severe. Some CAs are lethal in nature. Death from lethal anomalies can occur in-utero, at birth, during the perinatal period, or later at various stages of life. Others may have a normal span of life. The quality of life also is widely variable depending on the disorder.

4. Risk assessment of congenital anomalies

The assessment of risks for CA may not always be possible due to the lack of referable evidence and information. The known recurrence risks of some of the heritable disorders are presented in **Table 1**. Some of the risk factors for CA are well documented, such as parental age, subjects' gender, geographical location and exposure to drugs or toxins etc. Advanced maternal age is extensively reported to be associated with aneuploidies, such as trisomy 21, 13, and 18 and Klinefelter syndrome. Advanced maternal age has also been associated with non-chromosomal genitourinary anomalies,

Disease	Condition	Recurrence risk
Pyloric stenosis	Mother affected	19% risk for male, 7% for female offspring
	Father affected	5.5% risk for male, 2.4% for female
	1 child affected	4% risk to next male 2.4% to female child
Cleft lip	Unaffected parents, 1 child affected	4.5% risk for next child
	1 parent and 1 child affected	10% risk to next child
Cleft palate	1 child affected	2.6% risk to sibling
Congenital hip dysplasia	1 affected child	0.5% risk for male 6.3% for female child
Cardiac defects	1 child affected	3.4% risk for next child
	2 children affected	10% risk for next child
Neural tube defects	1 child affected	3.5% risk to next child
Trisomy 21	Mother with balanced translocation	10–15% risk for sibling
	Father with balanced translocation	5% risk for sibling
	1 child affected, no parental translocation	1% risk for sibling
Hirschsprung's disease	1 child affected	3–5% risk for next child
Club foot	1 affected child	2% risk if 1st child is male, 5% if female
	1 parent and 1 child affected	25% risk to next child

Table 1.
Recurrence risk of certain malformations. (adapted from Jones KL, Smiths recognizable patterns of human malformations, 5th edition, WB Saunders, Philadelphia, 1997; and Neonatology Review, 2nd edition, D. Brodsky & C. Martin, Hanley and Belfus Inc.2003.

as well as hips and feet deformities [13]. Advanced paternal age is associated with an increased incidence of de novo DNA mutations and chromosomal aberrations in the sperm, which may lead to miscarriage or genotypical and/or phenotypical anomalies in the fetus [14]. Like advanced maternal age, trisomy 21 is documented to be associated with advanced paternal age as well [15]. Other disorders reported to occur more commonly with advanced paternal age are achondroplasia, osteogenesis imperfecta, and some syndromes such as Apert, Waardenburg, Marfan, and Treacher Collins.

Green et al. have reported that the risks for cleft palate, diaphragmatic hernia, right ventricular outflow tract obstruction, pulmonary valve stenosis, anomalous pulmonary venous return, cataract, aortic coarctation, encephalocele, esophageal atresia, and multiple complex defects in the offsprings are enhanced with each unit year increase in the paternal age [14]. It is noteworthy that the effects of paternal age might vary in tandem with maternal age. While young maternal and paternal age are identified as independent risk factors for gastroschisis, young paternal age can become a risk factor for gastroschisis again if the mother's age exceeds 35 years. Omphalocele, spina bifida, orofacial clefts, and septal heart defects display associations with parental mating, which involves advanced paternal and young maternal age, and also between a young father and mother of advanced age.

Some diseases exhibit gender preferences [16, 17]. The diseases known to occur more commonly in males are pyloric stenosis, Hirschsprung's disease, imperforate anus, club foot, unilateral multicystic dysplastic kidney; cleft lip and palate, Poland sequence, ventricular septal defect, transposition of great vessels, aortic coarctation, hypoplastic left heart syndrome, subdiaphragmatic total anomalous pulmonary venous return, and pulmonic stenosis and atresia. Disorders identified to be more common in females are choanal atresia, choledochal cyst, congenital hip dysplasia, ureterocele, Trisomy 18, atrial septal defect, patent ductus arteriosus, anencephaly, meningomyelocele and congenital hypothyroidism. In a recent study of 12,795 cases with CA, male fetuses were found to be more susceptible to birth defects than females, however, with significant heterogeneity in the subtypes [16]. Sex organ anomalies are reported to be 8.5 times more common, and GIT defects 55%, whereas urinary tract anomalies 62% more prevalent in males than in females. Overall, the prevalence of major CA in males is 3.9% compared to 2.8% in females [17].

4.1 Calculation of carrier frequency and recurrence risk in a population

The Hardy-Weinberg equilibrium is utilized to predict carrier frequency in a given population and to calculate recurrence risk. The calculation assumes that the mating is random and there are no new mutations or natural selections.

4.1.1 Calculation of carrier frequency

Example: Normal gene frequency=P; Abnormal gene frequency =Q; P+Q =1 (i.e. 100%); 2PQ =carrier frequency; P^2= normal, non-carrier frequency; Q^2 = affected frequency.

Let us take the example of cystic fibrosis (CF), an autosomal recessive disease. The disease frequency of the morbidity is 1 in 2500 Caucasian births. I.e., Q^2 = 1/2500; therefore Q= 1/50. We know that P+Q=1. So P=1-Q or 1-1/50 = 49/50 ~ 1. Therefore 2PQ (carrier frequency) = 2 * 1 * 1/50 = 1/25.

4.1.2 Calculation of recurrence risk of CF

Example: a pregnant woman has a sister with CF. What are the chances of her having an affected child?

The Father's risk of being a carrier is 1/25, while the mother's chances of being a carrier are 2/3 as both of her parents are carriers. The chance of the offspring getting the gene from each parent is ¼. Therefore the chances of the child being affected by CF are 1/25X2/3X1/4=1/150.

5. Summary

Congenital anomalies are a heterogeneous group of disorders of abnormal morphogenesis, which present at birth and carry widely variable implications for morbidity and mortality. The basics of incidence, pathogenesis, and risk assessments of CA are discussed in this section.

Author details

Rita Prasad Verma
Nassau University Medical Center, East Meadow, NY, USA

*Address all correspondence to: rverma2@numc.edu

References

[1] Mburia-Mwalill A, Yang W. Birth Defects Surveillance in the United States: Challenges and Implications of International Classification of Diseases, Tenth Revision, Clinical Modification. Int Sch Res Notices. 2014; 2014: 212874. doi: 10.1155/2014/212874

[2] Hoyert DL, Xu JQ, "Deaths: preliminary data for 2011," National Vital Statistics Reports, vol. 61, no. 6, pp. 40-42, 2012.

[3] World Health Organization, "Congenital anomalies," *Fact Sheet* 370, World Health Organization, Geneva, Switzerland, 2014, http://www.who.int/mediacentre/factsheets/fs370/en/.

[4] Arth AC, Tinker SC, Simeone RM, Ailes EC, Cragan JD, Grosse SD. Inpatient hiospitalization costs associated with Birth defects among persons of all ages—United States 2013. MMWR Morb Mortal Wkly Rep 2017;66:41-46

[5] Queisser-Luft A, Stolz G, Wiesel A, Schlaefer K, Spranger J. Malformations in newborn: results based on 30,940 infants and fetuses from the Mainz congenital birth defect monitoring system (1990-1998). Arch Gynecol Obstet. 2002 Jul;266(3):163-7. doi: 10.1007/s00404-001-0265-4. PMID: 12197558.

[6] Hennekam RC, Biesecker LG, Allanson JE, Hall JG, Opitz JM, Temple IK, Carey JC, Elements of Morphology Consortium. 2013. Elements of morphology: General terms

[7] Prescott KR, Wilkie AO. Genetic aspects of birth defects: new understandings of old problems. *Arch Dis Child Fetal Neonatal Ed*. 2007;92(4):F308-F314. doi:10.1136/adc.2004.062968

[8] Chial, H. & Craig, J. (2008) mtDNA and mitochondrial diseases. *Nature Education* 1(1):217

[9] Rasmussen SA, Hernandez-Diaz S, Abdul-Rahman OA, et al. Assessment of congenital anomalies in infants born to pregnant women enrolled in clinical trials. *Clin Infect Dis*. 2014;59 Suppl 7(Suppl 7):S428-S436. doi:10.1093/cid/ciu738

[10] McDonald-McGinn DM, Emanuel BS, Zackai EH. 22q11.2 Deletion Syndrome. GeneReviews, Seattle, 2013. http://www.ncbi.nlm.nih.gov/books/NBK1523/ (Accessed on August 19, 2016)

[11] Schoumans J, Ruivenkamp C, Holmberg E, et al. Detection of chromosomal imbalances in children with idiopathic mental retardation by array based comparative genomic hybridisation (array-CGH). J Med Genet 2005; 42:699.

[12] Yang Y, Muzny DM, Xia F, et al. Molecular findings among patients referred for clinical whole-exome sequencing. JAMA 2014; 312:1870.

[13] Grossi, Matt MD; Crawford, Susan MSc; Chandra, Sue MD, FRCSC, MSc Advanced Maternal Age and Incidence of Non-Chromosomal Congenital Anomalies: A Population-based Study [35Q], Obstetrics & Gynecology: May 2017 - Volume 129 - Issue 5 - p 182S doi: 10.1097/01.AOG.0000514144.08234.ef.

[14] Green RF, Devine O, Crider KS, et al. Association of paternal age and risk for major congenital anomalies from the National Birth Defects Prevention Study, 1997 to 2004. *Ann Epidemiol*. 2010; 20(3):241-249. doi:10.1016/j.annepidem.2009.10.009.

[15] Jin Liang Zhu, Kreesten M. Madsen, Mogens Vestergaard, Anne V. Olesen, Olga Basso, Jørn Olsen, Paternal age and congenital malformations, *Human Reproduction*, Volume 20, Issue 11,

November 2005, Pages 3173-3177, https://doi.org/10.1093/humrep/dei186

[16] Tennant PW, Samarasekera SD, Pless-Mulloli T, Rankin J. Sex differences in the prevalence of congenital anomalies: a population-based study. Birth Defects Res A Clin Mol Teratol. 2011 Oct;91(10):894-901. doi: 10.1002/bdra.22846. Epub 2011 Aug 24. PMID: 21987467.

[17] Lary JM, Paulozzi LJ. Sex differences in the prevalence of human birth defects: a population-based study. Teratology. 2001 Nov;64(5):237-51. doi: 10.1002/tera.1070. PMID: 11745830.

Chapter 2

The Pathogenesis of Congenital Anomalies: Roles of Teratogens and Infections

Mehmet Semih Demirtaş

Abstract

Congenital anomalies present with significant financial, social, and moral issues and questions to the family and society and are difficult to rehabilitate. In utero exposure to teratogenic agents and infection are the two most important causes of nongenetic acquired anomalies presenting at birth. Teratogens such as drugs, adverse maternal conditions, and toxins are environmental factors that cause permanent structural or functional malformations or death of the embryo or fetus. Teratogens may cause significant congenital anomalies if encountered during the organogenesis period of 3–8 weeks of fetal life, which is the stage of tissues and organs formation, whereas minor morphological and functional disorders may occur with exposure during the fetal period of first 2 weeks. TORCH group infections (toxoplasmosis, others, rubella, cytomegalovirus, and herpes) are the most serious infectious diseases during pregnancy due to the severity of possible embryo-fetal lesions. With expanding scientific knowledge and clinical experience about the association of these toxins and infections with significant, at times crippling congenital anomalies, the avoidance of exposure to pregnant mothers has become the most important part of their prevention and management.

Keywords: teratogens, drugs, toxins, alcohol, smoking, congenital infections, cytomegalovirus

1. Introduction

In utero exposure to teratogenic agents and infection are the two most important causes of nongenetic, acquired anomalies presenting at birth. The fetal response and susceptibility to such agents are variable, and the effects depend on the type, timing, and duration of intrauterine exposure [1, 2] (**Figure 1**). The end results of such exposures may be organ system malformations; aberrations in organ growth, function, and development; and even death. The developmental stage of organogenesis, which is characterized by rapid cellular differentiation and migration, is the most vulnerable period, as the actively dividing cells are highly sensitive to the adverse effects of noxious agents [3]. The effects of teratogens during the preimplantation embryonic phase of the first two postconceptional weeks might present as all or none, as the uterine implantation of a defective embryo may fail and the pregnancy end with undetected abortion, thus nullifying the possibility of congenital malformations [4, 5] (**Figure 1**).

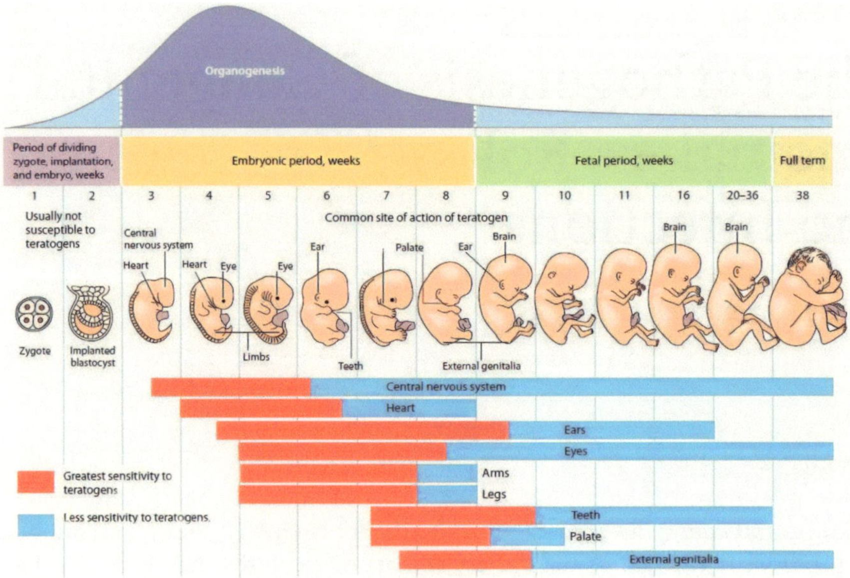

Figure 1.
Sensitivity to teratogens during pregnancy.

Congenital anomalies are health problems that are difficult to rehabilitate. They generate high treatment costs and might bring on huge financial and moral burdens to the family and society. According to the congenital anomalies survey conducted by the World Health Organization (WHO) in 193 countries in 2010, 270,000 of the 3.1 million newborn deaths were caused by congenital anomalies [6]. In the United States, 2–3% cases of the 3–5% of children born with birth defects are attributed to environmental or iatrogenic teratogen exposure during the intrauterine (IU) life [7]. Most of the teratogen-induced anomalies are preventable.

2. Teratogenic agents

Teratogens may cause significant congenital anomalies if encountered during the organogenesis period of 3–8 weeks of fetal life, which is the stage of tissue and organ formations (**Figure 1**). Minor morphological and functional disorders may occur with exposure during the fetal period of the first 2 weeks [8]. Multiple factors come into play for the teratogens to impart their effects. These are the genetic specifications of the conceptus, the dose and duration of exposure, and the mechanism of action of the offending agent. Teratogens effectuate primarily by disrupting cell-specific biochemical metabolism and by compromising blood circulation which lead to cell death. They can destroy and deplete essential nutrients, block enzyme activities, disrupt mitosis, interfere with nucleic acid functions, and derange membrane functions, osmolar balance, and energy production [9, 10]. Genetic differences in response to teratogens have been documented and may be due to the presence of genetic polymorphisms in the activities of enzymes involved in the excretion of toxic substances [11]. Animal studies have shown differences in the susceptibility to teratogen-induced damage within the same as well as between different species. Fetal hydantoin syndrome is detected in 5% of embryos exposed to phenytoin (PTN), and about 30% of them show some congenital anomalies, while more than half display no teratogenic effects [12]. Aspirin, corticosteroids, and some vitamins are teratogenic in mice and rats, but not in humans. Cleft palate and cleft lip are more common in mice with consanguineous matings [13].

2.1 Drugs

Drugs can directly affect the product of conception and cause malformation and/or embryo-fetal demise. They can impair the fetal development by compromising the transplacental transfer of nutrients and oxygen from the mother. They may diminish fetal blood supply and initiate premature myometrial contractions resulting in premature birth [14]. Drugs can play roles in the intrauterine development of gene-encoding proteins, thereby altering transcription regulation signals which adversely affect embryogenesis [15]. Drugs can exert their effects at different stages of cell development, namely, replication, proliferation, gene expression, signal transduction, programmed cell death, and cell migration (**Table 1**) [16, 17].

2.1.1 Phenytoin

Although the exact pathogenesis of phenytoin (PTN) embryo toxicity is unclear, some possible mechanisms have been proposed [18]. Phenytoin acts as a membrane stabilizer by inhibiting sodium (Na) and calcium (Ca) channels, as a result of which free radicals are released and cause endothelial damage, myocardial depression, bradycardia, and consequently fetal hypoxia. Phenytoin induces cytochrome P450 activation which results in the release of teratogenic free radicals, sourced via the metabolism of epoxides, folate, and vitamin K in the liver [19, 20]. Phenytoin, like other antiepileptic agents, namely, valproic acid (VPA) and vigabatrin, induces

Drug	Most susceptible period	Effects
Phenytoin	Organogenesis (18–60 days)	Fetal hydantoin syndrome, facial cleft, cognitive impairment
Lithium	Organogenesis (18–60 days)	Ebstein's anomaly
Warfarin	Second part of the first trimester (6–9 weeks)	Nasal hypoplasia, limb hypoplasia, optic atrophy, bone abnormalities, neurological impairment
Amphetamines	All trimester	Cleft palate, heart defects, intestinal atresias, and structural brain abnormalities
Sodium valproate	Organogenesis (18–60 days)	Neural tube defect, cleft palate, atrial septal defect, hypospadias, polydactyly, craniosynostosis
Cyclophosphamide	Organogenesis (18–60 days)	Skeletal and ocular defects, cleft palate
Aminopterin	Organogenesis (18–60 days)	CNS, limb, and skeletal defects
ACE inhibitors	Second. or third trimester (13th week term)	Craniofacial abnormalities, neonatal renal failure, pulmonary hypoplasia
Benzodiazepines	Organogenesis (18–60 days)	Cleft lift and palate abnormalities
Lithium	First trimester	Ebstein's anomaly

Table 1.
Some teratogenic drugs and their effects.

carnitine deficiency in the fetus which may lead to cardiomyopathies and ventricular septum defects [21]. Infants born to women with mutations in the methylenetetrahydrofolate reductase (MTHFR) gene are at an increased risk for fetal hydantoin syndrome as its protein products compromise the metabolism of phenytoin and/or its metabolites. Free radicals released as intermediate metabolites of phenytoin bind to deoxyribonucleic acid (DNA), proteins, and lipids and adversely affect the neurodevelopment. The wide variation in the presentation of anomalies related to PTN may be due to the genetic differences in the formation of free radicals, drug clearance, and repair mechanism. Fetal hydantoin syndrome can be seen in approximately 5–10% of infants with in utero exposure to phenytoin, whereas incomplete clinical syndrome can be seen in about one third of them [22]. The characteristic features of fetal hydantoin syndrome include microcephaly, craniofacial anomalies, hypertelorism, flattened nasal root, ptosis, wide mouth, cleft palate-lip, cardiac defects, urogenital malformations, and hypoplastic distal phalanx and nails. There is also an increased risk of neural tube defects (NTD) as this antiepileptic reduces fetal serum folate levels [23].

2.1.2 Valproic acid

Depending upon the dose and duration, the in utero exposure to VPA may increase the incidence of congenital malformations in neonates by 2–16 times [24]. The teratogenic effects of VPA on the fetus are typically caused by maternal ingestion of drug in doses over 1000 mg/day. However, adverse effects can be seen at lower doses of 500 mg/day as well. In one study, the rate of major congenital malformations with fetal exposure to VPA via maternal medication in the doses of <700 mg/day for 1 year was 6%, which increased to 10% when the doses were between 700 and 1500 mg and to 24% when over 1500 mg [25].

Like PTN the exact mechanism of action of VPA is unknown and various theories have been forwarded. Crudup et al. showed that VPA can increase γ-aminobutyric acid (GABA) levels in the brain via the inhibition of its catabolism [26]. VPA can directly inhibit voltage-gated sodium channels or bind to the proteins by acting as a histone deacetylase inhibitor (HDACi). HDACi can disrupt cell cycle, stop growth, and induce apoptosis [27, 28]. Furthermore, VPA induces chromatin changes and reduces the transcription of mRNA by converting chromatin segments to heterochromatin. The high affinity of valproic acid to folate receptors causes their competitive inhibition and increases the frequency of neural tube defects by as much as 20 times [29].

Valproic acid may cause multi-organ system anomalies, including those of craniofacies (epicanthal fold, small wide nose, anteverted nostril, long philtrum, thin upper-thick lower lip, retroverted ears), extremities (polydactyly, arachnodactyly, rudimentary fingers), and spinal column (neural tube defects, spina bifida). Other important defects include those of cardiovascular (ventricular septal defect (VSD), patent ductus arteriosus (PDA), aortic coarctation), respiratory (tracheomalacia), and urogenital systems (inguinal hernia, hypospadias, cryptorchidism, incomplete fusion of the Müllerian duct). The incidence of meningomyelocele, especially lumbar or lumbosacral, is reported to be 1–2% with in utero fetal exposure during the first trimester [30, 31]. Developmental anomalies and autism are other teratogenic effects of VPA described in the literature.

2.1.3 Thalidomide

Thalidomide (TD), which is currently being used for the treatment of multiple myeloma and leprosy, was initially prescribed for pregnancy-associated nausea and emesis in Europe, Asia, and America, without any preceding drug phase studies in the 1950s. Its teratogenic effects were first noticed in Europe in the early 1960s [32] when several case reports of phocomelia in babies born to mothers treated with TD during pregnancy were published. This revelation became a turning point in the history of pharmacovigilance. In November 1961, Lenz presented the specific features including limb anomalies in 52 in utero exposed infants to TD at a Pediatric Congress. This was followed by a publication documenting an association between the drug and congenital malformations in 1962 [33] (**Figure 2**). Subsequently, 115 case reports of similarly affected infants in Germany, Belgium, Sweden, and the UK were published, and the drug was withdrawn from the market [34]. Thalidomide affected about 10,000–15,000 infants and caused death in more than half of them during this period.

The incidence of congenital malformations is 50% if 50 mg of TD is ingested during the postfertilization days of 20–36 [35]. If given earlier it may cause miscarriages as demonstrated in rats. More than 30 scientific theories for thalidomide embryopathy have been forwarded over the past 50 years [36]. DNA mutagenesis, chondrogenesis, nerve/neural crest toxicity, and inhibition of cell adhesion molecules have been proposed as the potential mechanisms of thalidomide embryopathy. However, the most widely accepted theory is that of the antiangiogenic action of the drug on fetus [37]. D'Amato et al. showed that thalidomide inhibits angiogenic vascularization of the rodent cornea induced by a fibroblast growth factor protein. It is believed that thalidomide exerts its teratogenic effects by adversely affecting the embryonic blood vessels, which results in the disruption of vascularization during organogenesis leading to abnormal fetal growth and congenital malformations [38, 39]. The congenital anomalies caused by thalidomide are phocomelia, dementia, dysosmia, bone hypoplasia, cardiac malformations, ear malformations, splenic agenesis, gallbladder agenesis, and esophageal, duodenal, and anal atresia as well as stenosis [40].

2.2 Toxins

There has been a rapid progress in the awareness of adverse effects of a wide variety of environmental, medical, infectious, and nutritional toxins on the developing fetus since the end of the twentieth century. With the expanding scientific knowledge and clinical experience about the association of these toxins with significant, at times crippling, congenital anomalies, the avoidance of exposure to pregnant mothers has become the most important part of their prevention and management. The congenital malformations associated with exposure to the current known toxins

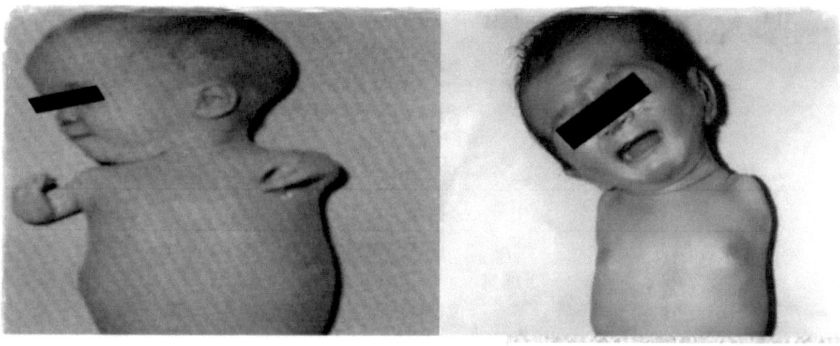

Figure 2.
Phocomelia and amelia.

are deafness, visual impairment, skeletal anomalies, and central nervous system (CNS) malformations, apart from embryonic loss and fetal demise [41].

2.2.1 Radiation

Radiation is a highly teratogenic toxic agent which exerts its adverse effects at cellular, subcellular, and molecular levels. It disrupts the molecular structure by both direct and indirect actions. No cell is known to be completely resistant to the toxicity of radiation. The risks are highest during the organogenesis phase [42], and the most vulnerable part of the cells to radiation injury is the highly active nucleus. The radiation-induced damage to the DNA may result in cell death, genetic mutations, and malformations, the severity and extent depending on the radiation dose and the stage of cell development at the time of exposure. Chromosomal anomalies are observed in cells when they are exposed to radiation during mitosis and DNA molecule formation [43, 44]. Cellular interruption and suppression of cell growth are the most common manifestations of radiation exposure during mitosis. Bergonie and Tribondeau (1906) documented that the most sensitive cells to radiation are the ones that are un- or underdifferentiated with undetermined function and morphology and are undergoing the highest mitotic activity [45]. The effects of radiation exposure during the first 14 days after fertilization are abnormal or failed embryo implantation resulting in miscarriage.

The dose is an important determinant of the radiation toxicity, and, accordingly, all pregnancies may not suffer from adverse effects [46]. As per the International Commission on Radiation Protection (ICRP), the chances of adverse or lethal effects in the preimplantation period of embryonic development are very low if the dose is less than 100 milliSieverts (mSv), and the actual threshold dose for the production of malformations is around 100 mSv [47]. The embryo is most susceptible to radiation-induced congenital malformations during the postconceptional ninth day and sixth weeks, the phase of organogenesis. Cerebral structural and functional anomalies such as microcephaly and mental retardation occur following exposure to doses over 100 mSv during the 8–16 weeks of intrauterine life, whereas ocular and skeletal abnormalities result with doses exceeding 200 mSv. After the sixth week of pregnancy and when the major part of organogenesis is competed, radiation causes neurodevelopmental delays. It is stated that the therapeutic risks of radiation are minimal in doses less than 50 mSv (**Table 2**) [48]. These dose-effect relationships were demonstrated in animal experiments. In humans, microcephaly and mental retardation were the most common anomalies identified in children exposed to radiation during early conception in Hiroshima and Nagasaki survivors

Gestational period (weeks)	Effects	Estimated dose amount
Preimplantation (0–2)	Miscarriage or is not affected	50–100 mSv
Organogenesis (2–8)	Congenital anomaly (skeletal system, genital, or eye)	200 mSv
8–15	Severe mental retardation (high risk)	60–130 mSv
8–15	Intellectual influence	Reduction of 25 intelligence coefficients per Sv
8–15	Microcephaly	200 mSv
16–25	Severe mental retardation (mild)	250–280 mSv

Table 2.
Effects of radiation doses according to the gestational age.

after World War II. Other anomalies noted were low birth weight, cataract, genital and skeletal malformations, and microphthalmos [49]. Streffer et al. suggested that after organogenesis, the effects of exposure may be similar to the postnatal effects with no major congenital anomalies encountered. They reiterated that the mammalian embryo and fetus are highly radiosensitive and the nature and sensitivity of induced biological effects depend upon the dose and developmental stage at irradiation [50].

2.2.2 Alcohol

Alcohol is an important teratogen with multisystemic adverse effects. No amount of consumption is safe during pregnancy. In the USA, one "standard" drink contains approximately 14 g of pure alcohol. This translates to 12 ounces of regular beer (5% alcohol), 5 ounces of wine (12% alcohol), and about 1.5 ounces of distilled spirits (40% alcohol). The 2016 National Institute on Alcohol Abuse and Alcoholism has defined prenatal alcohol exposure as follows: ≥6 drinks per week for ≥2 weeks or ≥3 drinks per occasion on ≥2 occasions, started at 3 months before pregnancy, or at diagnosis, and continued until delivery [51]. The fetus eliminates alcohol poorly at a rate of only 3–4% of the maternal rate. Moreover, part of the alcohol excreted via the fetal urine into the amniotic fluid is swallowed back, thus recirculating it into the system, and a small volume of amniotic fluid alcohol is absorbed into fetal compartments via a transmembranous route. These factors make fetus specifically more vulnerable to the adverse effects of maternal alcohol consumption [52].

As with other teratogenic agents, the effects of alcohol in the fetus vary according to the gestational age and the duration and dose of exposure [53]. Alcohol damages the structure, neuronal migration, and synaptogenesis in the developing CNS of the fetus. The consumption of two glasses of alcohol per day during pregnancy, especially the first 3 months, leads to the typical fetal alcohol spectrum disorder (FASD), which is characterized by structural, behavioral, emotional, and neurological problems in the offsprings [54]. The typical features of this syndrome are the minor facial anomalies, including short palpebral fissure, thinner upper lip, and flat philtrum. Significant pre- and postnatal growth retardation is a common feature, along with variable mental retardation which may manifest as a decrease in intelligence quotient, difficulties in perception, and delays in certain skills-seeking tasks (**Figure 3** and **Table 3**) [55, 56]. FASD may present with congenital cardiac defects as well, the most common being ventricular septal defect, atrial septal defect, conotruncal anomaly, and tetralogy of Fallot. The risk of conotruncal anomaly increases as per the amount of alcohol consumed during the periconceptional period [57].

2.2.3 Smoking and secondhand smoking

Cigarette smoking during pregnancy remains a major worldwide problem despite a significant decrease in incidence as a result of an increasing awareness of its adverse fetal effects. It is estimated that around 10–24% of women smoke while pregnant [58]. Fetal exposure to nicotine negatively affects its growth and increases the risk of neonatal and infant mortality and morbidity [59]. Nicotine and carbon monoxide (CO) decrease the placental blood flow via the vasoconstrictive effects of catecholamines, which are released from adrenals by nicotine activation. Nicotine promptly crosses the placental barrier and reaches its maximum activity in the fetus within 30 min of exposure. The concentration of nicotine in the amniotic fluid is demonstrated to be six times higher at 88% compared to 15%

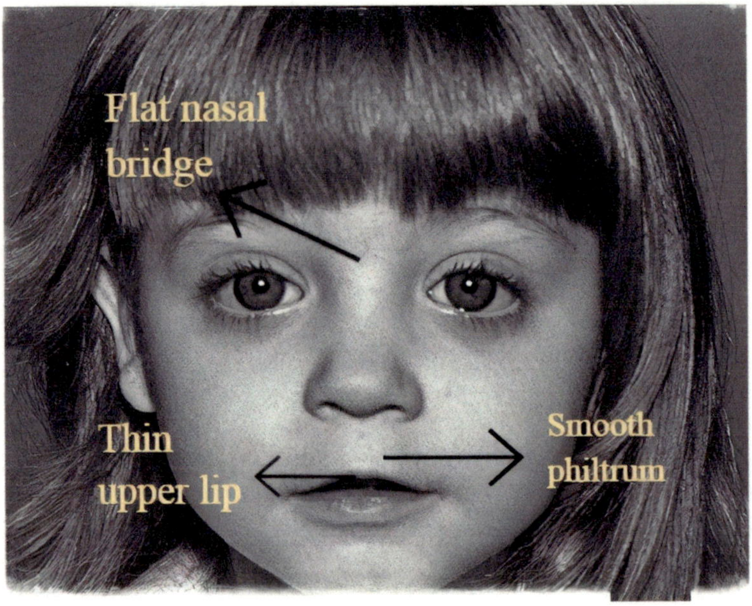

Figure 3.
Fetal alcohol syndrome (© 2009 University of Washington. With permission, Susan Astley, PhD).

1. Fetal alcohol syndrome (FAS) (all conditions will be met)
   A. Confirmation of alcohol use during pregnancy
   B. The presence of characteristic minor facial anomalies (at least two of the following):
      i. Short palpebral fissure (≤10th percentile)
      ii. Thin upper lip (score 4 or 5 in the lip/philtrum guide)
      iii. Straight philtrum (score 4 or 5 in the lip/philtrum guide)
   C. Prenatal and/or postnatal growth retardation:
      i. Length or weight ≤ 10th percentile
   D. Mental development disorder or abnormal morphogenesis (at least one of the following):
      i. Structural abnormalities in the brain
      ii. Head circumference ≤ 10th percentile
2. FAS in which alcohol use cannot be confirmed during pregnancy
3. Partial FAS in which alcohol use is confirmed during pregnancy
4. Partial FAS in which alcohol use cannot be confirmed during pregnancy
5. Alcohol-related congenital disorders
6. Alcohol-related neurodevelopmental disorders

Table 3.
Fetal alcohol spectrum disorders.

in the mother's blood. Nicotine acts on the brain by binding to nicotinic acetylcholine receptors (nAChRs) in autonomic ganglia and at neuromuscular junction. The binding results in the release of neurotransmitters and important neuromodulators, such as dopamine, adrenaline, acetylcholine, Seratonin (5- hydroxytryptamine), GABA, glutamate, and substance P [60].

Both nicotine and carbon monoxide induce degenerative changes and premature aging in the placenta. The degenerative changes are marked by an increased amount of collagen in the chorionic villi and the thickening of subtrophoblastic basement membrane [61]. Premature aging is suggested by the increase in the syncope buds and apoptosis in the placentas. Both premature aging and degenerative changes significantly reduce the placental functional capacity and lead to multiple adverse fetal effects. The incidence of premature births is significantly higher in mothers who smoke [62]. The results of a recent meta-analysis by Hackshaw et al. demonstrated that maternal smoking increases the risks for a variety of system malformations, including those of cardiovascular (cardiac septal defects, malformations of pulmonary and tricuspid valves, and malformations of the great arteries), musculoskeletal (limb reduction, clubfoot), craniofacial (craniosynostosis, cleft lip and palate), and gastrointestinal (gastroschisis) [63].

2.3 Adverse maternal conditions: nutritional deficiencies, diseases, and infections

Nutrients taken during pregnancy can have significant and lasting effects on maternal and newborn health. Inadequate or excessive intake of nutrients if associated with consequent pathophysiological changes during pregnancy can bring about epigenetic changes in the fetus with adverse short- and long-term implications. Optimum intakes of energy and nutrients during pregnancy as well as during breastfeeding are essential for the initiation and maintenance of a healthy life during childhood. It may also protect against several adulthood diseases [9, 64].

2.3.1 Zinc

Zinc is essential for normal fetal growth and development. It is a component of over 200 enzymes which take part in the formation and release of various proteins, hormones, and neuropeptides. The element is involved in the transcription process in which a gene's DNA sequence is copied to make an RNA molecule. Zinc is required for proper cell division, growth, and differentiation. Severe zinc deficiency is embryotoxic and teratogenous and may cause lethal fetal developmental and structural anomalies [65].

It has been shown that maternal zinc deficiency can affect embryonic protein and DNA synthesis and cause chromosomal damage characterized by terminal deletion [66]. Maternal zinc deficiency is associated with increased apoptosis in the embryonic cells. TUNEL analysis has shown that cell death is increased in the peri-implantation embryos if the cultured cells have low zinc levels [67]. The cell cycle may not be adversely affected if the maternal zinc deficiency is short term [68]. In zinc deficiency, the formation of free radicals is increased as they cannot bind to the membranes and intracellular regions of redox-active metals, such as copper and iron [69], which results in increased oxidative stress and teratogenicity. Zinc can prevent oxidation of numerous proteins, including zinc finger transcription factors of redox-sensitive cysteine and sulfhydryl groups. Zinc is a component of copper-zinc superoxide dismutase and is the regulator of metallothionein, a metal-binding protein which has important roles in the execution of various physiological processes and in the prevention of stress [70]. Metallothionein releases zinc, which plays a central role in the antioxidant defense system during oxidative stress. Inadequate zinc uptake of the mother leads to a decrease in the circulating zinc levels which may adversely affect the neural tube development of the fetus as demonstrated in both animals and humans. In humans, the risk of neural tube defects is found to be increased in women with acrodermatitis enteropathica, a rare genetic disorder of zinc metabolism. It is noted that the prevalence of neural tube defects is higher in Africa and the Middle East, where zinc intake is chronically low due to ecological reasons [70, 71]. The relationship between zinc deficiency and cleft palate and lip was demonstrated in a study which showed the zinc levels in the blood of such infants and their mothers to be significantly low [72].

2.3.2 Folic acid (FA)

Folic acid, a group B vitamin, plays an important role in the production of new cells by assisting in the production of DNA and RNA that control cell proliferation [73]. It also works with vitamin B12 to form hemoglobin in erythrocytes. It has a protective effect against heart diseases. It decreases the risk of birth of infants with neural tube defects (spina bifida), obstructive urinary tract anomalies, limb deficiencies, orofacial clefts, and congenital hypertrophic pyloric stenosis [74]. After absorption, folic acid is carried as a monoglutamate in the blood and is converted to various compounds in the cell, the most important being the reductase enzyme, tetrahydrofolate (THF). THF functions as the donor of single carbon units at various steps of DNA synthesis, which is required for the synthesis of purines, thymidylate and hence thymine [75, 76].

Research on the effect of folic acid on NTD began in the 1980s when studies showed that FA is effective in preventing both primary NTD and its recurrence [77]. In a multicenter randomized controlled study which included 1200 women with a history of NTD in their prior pregnancies, FA intake in the dose of 0.4 mg/day started at least 1 month before conception and continued during the first 3 months of pregnancy reduced the risk of NTD by 3.6 times [78]. In a cohort study in China, which included approximately 250,000 women, it was demonstrated that maternal intake of 0.4 mg folic acid reduces the risk of NTD in the fetus by 85% in high-prevalence areas and by 40% in low-prevalence areas [79, 80]. Folic acid has been also reported to reduce the incidence of CHD if used during the preconceptional period [81]. The use of folic acid antagonist drugs, which cause the inhibition of dihydrofolate reductase enzyme, increases the frequency of CHD.

2.3.3 Maternal diabetes mellitus

The discovery of insulin in 1922 and advances in obstetrics and neonatal intensive care reduced perinatal mortality in pregnancies complicated by diabetes mellitus by approximately 30 times. By maintaining maternal euglycemia, such pregnancies were able to continue until term with a resultant decrease in prematurity-related complications including respiratory distress syndrome [82]. Still, perinatal mortality in diabetic women continues to be about twice that of nondiabetic women. Also spontaneous abortion rates are higher in diabetic women, especially if the glycemic control is suboptimal in the periconceptional period [83].

Hyperglycemia has been shown to induce oxidative stress in the developing embryonic and fetal cells and tissues in animal studies, with the release of reactive oxygen species (ROS). Increased concentrations of ROS induce organ malformation and birth defects via membrane changes, mitochondrial dysfunction, and the initiation of abnormally programmed cell death (apoptosis). In mice models injected streptozotocin (STZ) to induce type 1 diabetes hyperglycemia caused changes in the yolk sac, as well as abnormalities in the endoplasmic reticulum and premature aging. It induced oxidative phosphorylation in the mitochondria and increased the concentration of ROS [84]. If appropriate glycemic control is maintained during the third and sixth weeks of pregnancy, the periods when the embryo is most susceptible to teratogens, congenital anomaly rates are found to be the same as in the general population [85]. Vitamins E and C, which are antioxidants, have been shown to reduce hyperglycemia-related anomalies in animal models. Some prostaglandins may have the same effects.

The incidence of congenital anomalies, which is 1–2% in the general population, is 4–8 times higher in infants of mothers with pregestational diabetes. Congenital anomalies are the most important cause of perinatal death in pregnancies complicated with diabetes mellitus [86]. Although anomalies can be seen in all organ systems in the neonates of diabetic mothers, the most important ones are those in the cardiac and central nervous systems (**Table 4**). Caudal regression syndrome is a rare congenital anomaly caused by maternal diabetes. No increase in the rate of congenital anomalies is seen in normoglycemic mothers or those with gestational diabetes occurring after the first trimester, which reiterates that glycemic control during embryogenesis plays a major role in the pathogenesis of fetal anomalies. Congenital anomalies are found to be more common in pregnant women with high HbA1c levels in the first trimester with a direct relationship with its level and the rate of anomalies [87].

2.3.4 Maternal phenylketonuria (PKU)

Maternal phenylketonuria is one of the most common teratogenic syndromes of pregnancy. Phenylalanine crosses the placenta by active transport and increases the level of phenylalanine in fetal blood by 70–80% of maternal phenylalanine concentration [88]. Increased levels of phenylalanine are toxic and teratogenic to the developing fetus. Spontaneous abortions are observed in 24% of pregnancies with phenylketonuria, and in those who survive, microcephaly is found in 73%, mental retardation in 92%, congenital heart diseases in 12%, and intrauterine growth retardation in 40% of the offsprings [89]. If maternal phenylalanine levels are well controlled before conception and during pregnancy, the incidences of microcephaly and abnormal physical and neurological fetal development are significantly reduced. The prognosis is best in infants of mothers with a blood phenylalanine level of 120–360 μmol/L prior to pregnancy with no increase in the risks, while the prognosis is poor in those infants whose mothers had a phenylalanine level exceeding 360 μmol/L during pregnancy. Severe congenital heart diseases were reported in infants born to untreated pregnant women with high blood phenylalanine levels, especially if the diet restriction was not started until the 7th and 18th weeks of gestation. There is no increase in the risk in pregnant women with phenylalanine level 120–360 μmol/L during the first 8 weeks of pregnancy. Serious fluctuations in maternal phenylalanine levels in pregnancy also have a negative impact on prognosis [90].

In pregnant women with phenylketonuria, sapropterin dihydrochloride, an orally active synthetic form of (6R)-L-erythro-5,6,7,8-tetrahydrobiopterin, has been used in the doses of up to 20 mg/kg/day, in combination with a restricted diet for therapy, and the short-term results have been good. Large neutral amino acid

• Cardiac anomalies	• Atrial septal defect
• Central nervous system anomalies	• Ventricular septal defect
• Renal anomalies	• Transposition of large vessels
• Gastrointestinal abnormalities	• Aortic coarctation
• Neural tube defect	• Fallot tetralogy
• Anencephaly	• Trunkus arteriosus
• Uretral duplication	• Dextrocardia/cardiomegaly
• Duodenal-anorectal atresia	• Caudal regression syndrome
	• Sacral agenesis
	• Omphalocele

Table 4.
Congenital anomalies seen in children of diabetic mothers.

(LNAA) treatment, which is one of the other dietary alternatives used in patients with phenylketonuria, is contraindicated in pregnancy because it does not reduce blood phenylalanine levels to safe levels [91].

3. Maternal infections

Congenital anomalies caused by intrauterine exposure to infectious pathogens, especially certain viruses, continue to be a significant clinical problem around the world, despite the availability of vaccines (effective against rubella, varicella-zoster, and hepatitis B viruses), drugs (against herpes, toxoplasma, and HIV), and specific and sensitive immunological diagnostic tests for the majority of them. With the help of highly sensitive diagnostic procedures, the incidence of intrauterine infections during pregnancy is estimated to be about 12–20%. These infections cause a wide range of major anomalies and dysfunctions, including deafness, blindness, neurodevelopmental aberrations, growth failure, and congenital cardiac defects [92]. These diseases have been traditionally dealt with under the title of TORCH complex [93] representing toxoplasmosis; other (syphilis, parvovirus); R, rubella (German measles); C, cytomegalovirus; H, herpes simplex virus (**Table 5**).

The fetus and embryo are highly susceptible to infections, especially during the organogenesis period, while those encountered earlier may end in abortion. The fetus does not synthesize IgG and cannot adequately synthesize IgM and IgA until the second half of pregnancy. It has a very poor cellular immunity and the production of the necessary cytokines is suboptimal [94]. Some pathogens may infect the mother and the placenta without showing any clinical symptoms in mother and lead to miscarriage, congenital anomalies, preterm birth, fetal hydrops, and premature rupture of the membrane.

3.1 Toxoplasmosis

Congenital toxoplasmosis occurs due to the transplacental passage of acute maternal infection with the protozoan organism *Toxoplasma gondii* to the fetus. In neonates, it presents with a wide spectrum of clinicopathological features. It may be clinically asymptomatic and present in a serologically detected form only, at times, to manifest clinically only in later years. On the other end, it may display severe multisystem involvement with debilitating features, such as chorioretinitis, hydrocephalus, and intracranial calcifications [95]. The severity of fetal infection depends on the gestational age during the parasitemia. In early pregnancy, the placenta is an effective barrier to the parasite. The risk of congenital infection increases with increasing gestational age, the incidence being 15–17% in the first, 25% in the second, and 65% in the third trimester [96]. However, the severity and sequelae of the infection are much higher if encountered during early pregnancy. The incidence

• Cytomegalovirus	• Varicella-zoster virus (VZV)
• Human immunodeficiency virus (HIV)	• *Trypanosoma cruzi* (Cong. Chagas disease)
• Herpes simplex virus (HSV)	• Parvovirus
• *Toxoplasma gondii* (Cong. toxoplasmosis)	• Rubella
• Zika virus	• *Treponema pallidum* (Cong. syphilis)

Table 5.
Some strains of infection causing congonse to teratogens are very diverse and depend on the genetic sensitivity and severity of exposure.

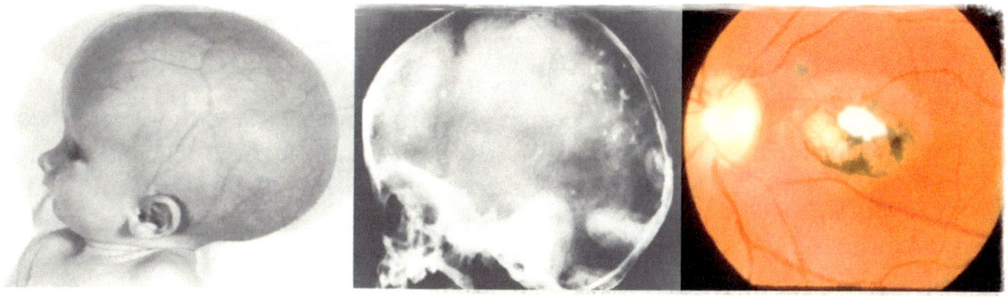

Figure 4.
Hydrocephalus, chorioretinitis, and intracerebral calcification findings in congenital toxoplasmosis.

of organ anomalies is 75% with the fetal infection in the first trimester and 5% if the infection happens during the third trimester. Overall, organ anomalies in congenital toxoplasmosis can be detected in 10–20% of infected fetuses. Preterm birth and intrauterine growth retardation are other complications of intrauterine toxoplasmosis [97]. Fetal infection during the first trimester of pregnancy may result in miscarriage or otherwise present as congenital infection with organ abnormalities specific to toxoplasmosis. Fetal infection occurring in the third trimester of pregnancy is often mild and asymptomatic depending upon the maturation of the fetal immune system and may at times only be serologically detected [98]. In addition, clinical symptoms may appear months or even years after birth. In a prospective study, it was reported that visual disturbances developed and detected during the regular well-child visit examinations of newborns asymptomatic at birth were diagnosed by serological to be due to toxoplasma by age as late as 20 years.

Other features of congenital toxoplasmosis are hydrocephalus, corpus callosum agenesis, cerebral calcification, microcephaly, intrauterine growth retardation, and nonimmune hydrops fetalis (**Figure 4**) [96, 99]. Pathologically, the placenta becomes first infected and appears to be pale, sludge-like, and edematous. Placental vasculitis along with granulomatous inflammatory lesions characterized by polymorphonuclear and lymphocyte infiltration in chorionic villi is specific for the disease. Following fetal transmission, fetal vasculitis develops [100], and the spreading trophozoites tend to settle in the brain and eyes. They form a granulomatous infectious lesion in the brain and its membranes. Eventually tissue necrosis happens around the parasite followed by fibrosis. These pathological processes cause congenital toxoplasmosis-specific microcephaly, cerebral calcifications, hydrocephalus, and chorioretinitis. It is reported that brain damage is more prominent in fetal infection before the 18th week of gestation. Eyes are the most commonly involved organ in congenital toxoplasmosis in which melanin pigment distribution disorder in the uvea and yellow-white edematous retinitis in the retina can be detected [101]. Microphthalmia and optic nerve atrophy may develop in infection encountered during early pregnancy. Deafness may develop in infants due to internal ear involvement. Skeletal muscles and myocardial infections are frequently involved. Moderate pneumonitis emulating viral pneumonia can be detected in the lungs. The liver usually enlarges and may present with pathological changes such as bile stasis, extramedullary hematopoiesis, dystrophic calcification, and portal fibrosis, while the pancreas, genital organs, urinary system, and gastrointestinal organs are generally not affected by the organism. The lymphoid tissue is affected and splenomegaly and adenopathy are commonly seen [102]. Clinically, fever, jaundice, espy direct, respiratory distress due to lung involvement, cardiovascular compromise due to myocarditis, hydrocephalus, and at times convulsions may be seen during neonatal period and in later life. In childhood, retardation is evident in mental and physical development. Vision and hearing disorders may occur in later life [103].

3.2 Rubella

Despite the availability of an effective vaccine, over 100,000 cases of congenital rubella syndrome are reported every year. Congenital rubella infection occurs via transplacental transmission of the RNA virus to the fetus during maternal viremia [104]. The infection may be asymptomatic, present with mild common cold-like symptoms, or devastating. Clinical features such as fetal-neonatal cataract and glaucoma, microphthalmia; patent ductus arteriosus, cardiac septal defects, pulmonary vascular stenosis, sensorineural deafness, fetal growth restriction, thrombocytopenia, anemia, hepatosplenomegaly, hepatitis, direct hyperbilirubinemia, chronic diffuse interstitial pneumonia, osseous changes, and even chromosomal abnormalities are grouped in congenital rubella syndrome. Fetal infection is found to cause necrosis, apoptosis, and division errors of cells involved in organogenesis, resulting in malformations. The mitotic activity is noted to be reduced in rubella-infected cells. Another potential mechanism is that of a direct viral invasion into vasculatures causing tissue necrosis without inflammation (**Figure 5**) [105]. The capsid of the rubella virion (RV) plays an important role in mitochondrial damage and viral replication complexes, as evidenced by immunogold electron microscopy and indirect immunofluorescence studies. Cardiolipin is a phospholipid associated exclusively with mitochondria, and its presence in rubella virions suggests the involvement of the internal mitochondrial membrane of cells in viral proliferation. Both the mitochondrial distribution and morphology are abnormal in RV-infected cells, and the mitochondria tend to cluster in the perinuclear region along with viral replication complexes. In advanced infection electron-dense plaques between opposing mitochondria are formed, and the mitochondrial cristae may be lost in RV-infected cells [106]. The risk of congenital malformation is low after 17 weeks of IU life. It is thought that the immune response mechanisms (immunoglobulins like IgM, IgG, and IgA, T cells, natural killer cells, and interferons) appear during the second half of pregnancy and are not sufficiently mounted against the infection during the first trimester [107].

3.3 Parvovirus

Parvovirus, a non-enveloped single-stranded DNA virus, is the only member of the *Parvoviridae* family known to cause human disease. Failure to produce the virus in cell culture has made it difficult to elucidate the pathophysiology. The symptoms of parvovirus B19-related infection are usually related to the host's immunological and hematological status. Since the immunological functions of the fetus are not optimum, parvovirus B19 infection may cause intrauterine infection, presenting as fetal anemia, hydrops fetalis, congenital malformations, and at times fetal-neonatal death [108].

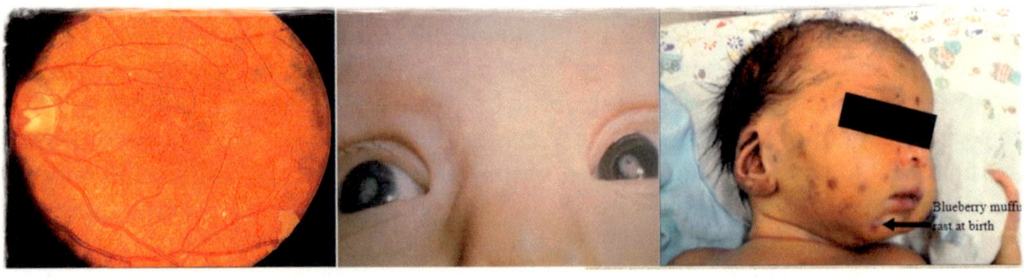

Figure 5.
Chorioretinitis, cataract, and blueberry muffin skin rash due to congenital rubella.

The vertical transmission of acute infection during pregnancy happens in 17–33% of the cases of maternal infection. Intrauterine infection is often asymptomatic. The fetus is highly susceptible to erythroid hypoplasia due to parvovirus B19 infection, and due to the shorter life span of fetal erythrocytes and the destructive effect of the virus, especially on rapidly proliferating erythrocytes, severe aplastic anemia may develop with the consequent hydrops fetalis. In addition to anemia, thrombocytopenia, leukopenia, elevated transaminase, and increased bilirubin, espy direct may also occur. Intrauterine parvovirus infections may also cause central nervous system anomalies. Fetal loss rate due to parvovirus B19 infection in pregnancy has been reported to be 3–9% and that of hydrops fetalis as 18% [109]. Overall the short-term prognosis of neonates with intrauterine parvovirus B19 infections is reported to be good.

If parvovirus B19 is diagnosed during pregnancy, serial fetal monitoring by Doppler ultrasonography to measure the fetal middle cerebral artery flow velocity in order to evaluate the need of intrauterine fetal blood transfusion is important. The mortality is reduced from 50 to 18% with intrauterine erythrocyte transfusion in conditions of hydrops and/or anemia detected antenatally. One transfusion is often sufficient [110]. The infection usually does not cause intrauterine death if appropriate transfusion and other supportive treatments are provided, and the fetus if born alive has good prognosis [109, 111]. Currently, no specific antiviral agent or vaccine is available for parvovirus B19.

3.4 Cytomegalovirus (CMV)

CMV is the most common organism causing congenital infection around the world. The frequency is reported as 5–12/1000 live births. The risk of transmission increases with increasing gestational age, but the severity of fetal disease decreases. Approximately 10–15% of the fetuses infected in early pregnancy are symptomatic at birth, and in those cases severe systemic invasive disease marked by intrauterine growth restriction, hepatosplenomegaly, cholestasis, transaminitis, abnormal liver function tests, pneumonia, pancytopenia, hemolytic anemia, petechia, purpura, and central nervous system anomalies is noted [112]. Central nervous system findings in congenital CMV infection are quite diverse. Microcephaly, sensorineural hearing loss, chorioretinitis, and convulsion are the most common presentations. Abnormal neuroradiological findings, including ventriculomegaly-hydrocephalus, increased periventricular echogenicity and calcification, white matter involvement, and lenticulostriate vasculopathy, are detected in 70% of the cases. Neuronal migration anomalies, temporal cystic periventricular leukomalacia, occipital intraventricular septa, cerebral atrophy, corpus callosum dysgenesis, and cerebellar hypoplasia are other relatively uncommon findings [113].

The inflammation process in the placenta infected with CMV is characterized by ICAM-1 expression on the membranes of placental trophoblasts, with enhancement in the adhesion of maternal blood cells [114]. During the mother's primary infection, virus-bearing infectious leukocytes transmit CMV infection to the trophoblasts, and through the trophoblasts, the CMV reaches the stromal fibroblasts and fetal endothelial capillary cells [115]. Further in the process the virus is directed toward and proliferates in the major target fetal organs, namely, the brain, liver, inner ear, spinal cord, kidney, and the vascular epithelium. Viral DNA replication takes place in the infected organs with the production of infectious viral progeny. Neurons, oligodendroglia, microglia/macrophages, and neural progenitor/stem cells, especially astrocytes, are particularly predisposed to CMV infections and may act as hosts in the replication and assist in the spread of the virus. The activated apoptosis during organogenesis is the important mechanism that leads

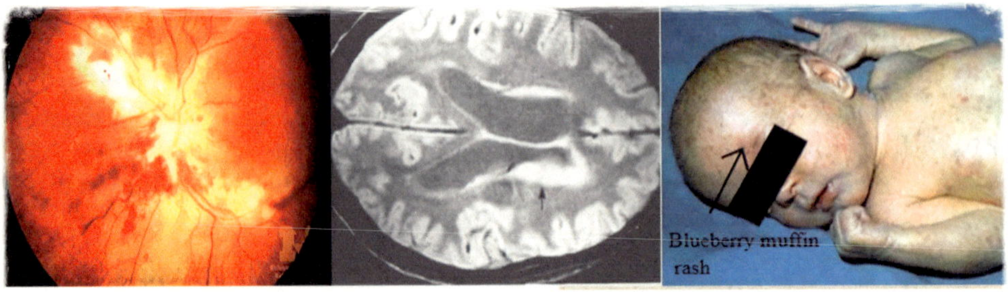

Figure 6.
Congenital CMV-induced chorioretinitis, intracerebral calcifications, and blueberry muffin skin rash.

to malformations [114, 116]. The sensitivity of CMV IgM test, which is frequently used for the diagnosis of congenital CMV infection, is low, and the false positivity rate is high. Urine and saliva cultures for the virus are the recommended investigations for the identification of infection (**Figure 6**) [113, 116].

The Infectious Diseases Committee and the American Academy of Pediatrics recommend that ganciclovir treatment be considered in patients with congenital CMV infection with symptomatic central nervous system involvement (microcephaly, intracranial calcification, hearing impairment, and retinitis). Ganciclovir is a deoxyguanosine analogue and the first antiviral drug shown to be effective in the treatment of CMV infection in humans. It is first phosphorylated to ganciclovir monophosphate by a viral kinase encoded by the CMV gene UL97 during infection. Then cellular kinases catalyze the formation of ganciclovir diphosphate, and ganciclovir triphosphate is a competitive inhibitor of deoxyguanosine triphosphate incorporation into DNA and preferentially inhibits viral DNA polymerases. Ganciclovir triphosphate inhibits the binding of deoxyguanosine triphosphate to viral DNA, slows viral DNA chain construction, and forms noninfectious viral DNA fragments. The concentration of ganciclovir triphosphate in infected cells is 10 times that of uninfected cells with a half-life in the cell longer than 24 hours [117]. Ganciclovir triphosphate also serves as a poor substrate for chain elongation, thereby disrupting viral DNA synthesis via a second route. However, clinically, ganciclovir treatment remains controversial in congenital CMV infection due to the need of long-term intravenous therapy, frequency of side effects, and limited healing from the infection. The use of valganciclovir, the L-valyl ester of ganciclovir, which is rapidly metabolized to ganciclovir in the body after oral administration, is increasing as with this treatment the need of parenteral therapy, hospitalization, and the risk of catheter-related infection are eliminated. The antiviral therapy may reduce the risk and duration of hospitalization in infants and is also reported to have a positive long-term effect on hearing [117, 118].

3.5 Varicella-zoster

The incidence of varicella infection in pregnancy is approximately 0.4–2.4/1000. The infection can result in severe fetomaternal complications. Spontaneous abortion with varicella infection is observed in the first trimester [119]. Congenital varicella syndrome occurs secondary to infection in the first two trimesters, and the infection is thought to result from the reactivation of varicella and invasion of the placenta, similar to the mechanism of herpes zoster. The syndrome generally presents with an abnormal development of musculoskeletal system, dermatomal pattern of skin lesions, and segmental dysfunction of somatic and autonomic nervous systems [120]. About 2% of fetuses exposed to the virus during the first 20 weeks of pregnancy (particularly during the 6th–20th week of gestation) may develop congenital varicella syndrome if the mother had no prior exposure to varicella. If varicella-zoster virus (VZV) infection occurs later during pregnancy (i.e., in the middle of the second or in the third trimester), the fetal immune system may be able to mount a response to the invading organism, typically resulting in a benign course. Embryopathy is not reported after 28 weeks. In one study, the incidence of varicella syndrome due to varicella infection during the first 20 weeks of pregnancy was reported as 0.91%, and the syndrome was not observed after 28 weeks [121]. In congenital varicella syndrome, the frequencies of occurrences of various systemic anomalies are as follows: skin lesions approximately 70%; limb hypoplasia 46–72%; nervous system abnormalities, such as cortical atrophy, microcephaly, and mental retardation, 48–62%; eye anomalies such as microphthalmia, cataract, and chorioretinitis 44–52%; and muscle hypoplasia, gastrointestinal, genitourinary, and cardiovascular system abnormalities, and developmental delay 7–24% [122].

3.6 Zika virus

In 2016, the US Centers for Disease Control and Prevention pronounced Zika virus infection as a risk for severe CNS defects in the fetuses of infected mothers. After crossing the placenta, the virus proliferates in the fetal brain tissues and infects the progenitor neural cells, leading to the growth failure and death of neural cells [123]. Although very few cases of Zika embryopathy are reported, the Zika virus-related CNS abnormalities are noted to be as follows: microcephaly, ventriculomegaly, cerebral calcifications, absent corpus callosum, and atrophy of the cerebellum and brainstem.

4. Summary

Congenital anomalies present with significant financial, social, and moral issues and questions to the family and society and are difficult to rehabilitate. In utero exposure to teratogenic agents and infection are the two most important causes of nongenetic, acquired anomalies presenting at birth. Teratogens are environmental and other agents that can cause structural or functional anomalies, or even demise in the embryo or fetus. TORCH (toxoplasmosis, others, rubella, cytomegalovirus, herpes) and other more recently identified infections during pregnancy may present with embryo-fetal systemic lesions of varying severity and result in significant morbidity and mortality. Most of the teratogen-induced and several infection-associated anomalies are preventable. Multiple factors determine the occurrence, presentation, and severity of congenital malformations in neonates who are exposed in utero to teratogens or infections. The individual response to teratogens is very diverse and depends on the genetic sensitivity of the product of conception and the severity of exposure.

Author details

Mehmet Semih Demirtaş
Department of Pediatrics, Aksaray University Education and Research Hospital, Aksaray, Turkey

*Address all correspondence to: md.semihdemirtas@gmail.com

References

[1] Hennekam RC, Biesecker LG, Allanson JE, Hall JG, Opitz JM, Temple IK, et al. Elements of morphology: General terms for congenital anomalies. American Journal of Medical Genetics: Part A. 2013;**161A**:2726-2733

[2] Lenz W, Knapp K. Foetal malformations due to thalidomide. German Medical Monthly. 1962;7:253-258

[3] Luteijn JM, Dolk H, Addor MC, Arriola L, Barisic I, Bianchi F, et al. Seasonality of congenital anomalies in Europe. Birth Defects Research. Part A, Clinical and Molecular Teratology. 2014 Apr;**100**(4):260-269. DOI: 10.1002/bdra.23231

[4] Maitra A. Genetic and pediatric diseases. In: Kumar V, Abbas AK, Aster JC, editors. Robbins Basic Pathology. 9th ed. Philadelphia: Elsevier; 2013. pp. 243-298

[5] Moore KL, Persaud TVN, Torche MG. The Developing Human: Clinically Oriented Embryology. 10th ed. Philadelphia: Elsevier; 2015

[6] WHO/CDC/ICBDSR. Birth Defects Surveillance: A Manual for Programme Managers. Geneva: World Health Organization; 2014

[7] Finnell R. Teratology: General considerations and principles. The Journal of Allergy and Clinical Immunology. 1999;**103**:S337-S342. DOI: 10.1016/S0091-6749(99)70259-9

[8] Cassina M, Cagnoli GA, Zuccarello D, Di Gianantonio E, Clementi M. Human teratogens and genetic phenocopies. Understanding pathogenesis through human genes mutation. European Journal of Medical Genetics. 2017;**60**(1):22-31. DOI: 10.1016/j.ejmg.2016.09.011

[9] Alexander PG, Clark KL, Tuan RS. Prenatal exposure to environmental factors and congenital limb defects. Birth Defects Research. Part C, Embryo Today. 2016;**108**(3):243-273. DOI: 10.1002/bdrc.21140

[10] Calado AM, Dos Anjos Pires M. An overview of teratology. Methods in Molecular Biology. 2018;**1797**:3-32

[11] Brent RL. Environmental causes of human congenital malformations: The pediatrician's role in dealing with these complex clinical problems caused by a multiplicity of environmental and genetic factors. Pediatrics. 2004;**113**:957-968

[12] Jones KL, Jones MC, del Campo Casanelles. Fetal hydantoin syndrome (fetal dilantin syndrome). In: Smith's Recognizable Patterns of Human Malformation. 7th ed. Philadelphia, PA: Elsevier Saunders; 2013. p. 734

[13] Niebyl J, Simpson J. Teratology and drugs in pregnancy. In: Global Library of Women's Medicine; 2008. DOI: 10.3843/GLOWM.10096

[14] Van Rooij IA, Miller RK, Zielhuis GA, de Jong-van den Berg LT, Roeleveld N. Teratogenic mechanisms of medical drugs. Human Reproduction Update. 2010;**16**(4):378-394. DOI: 10.1093/humupd/dmp052

[15] Thorpe PG, Gilboa SM, Hernandez-Diaz S, Lind J, Cragan JD, Briggs G, et al. Medications in the first trimester of pregnancy: Most common exposures and critical gaps in understanding fetal risk. Pharmacoepidemiology and Drug Safety. 2013;**22**:1013-1018

[16] Van Gelder MMHJ, de Jong-van den Berg LTW, Roeleveld N. Drugs associated with teratogenic mechanisms. Part II: A literature review

of the evidence on human risks. Human Reproduction. 2014;**29**:168-183

[17] Kennedy DS. A to X: The problem of categorisation of drugs in pregnancy—An Australian perspective. The Medical Journal of Australia. 2011;**195**:572-574

[18] Webster WS, Howe AM, Abela D, Oakes DJ. The relationship between cleft lip, maxillary hypoplasia, hypoxia and phenytoin. Current Pharmaceutical Design. 2006;**12**:1431-1448

[19] Nilsson MF, Ritchie H, Webster WS. The effect on rat embryonic heart rate of Na^+, K^+, and Ca^{2+} channel blockers, and the human teratogen phenytoin, changes with gestational age. Birth Defects Research. Part B, Developmental and Reproductive Toxicology. 2013;**98**:416-427

[20] Azarbayjani F, Danielsson BR. Phenytoin-induced cleft palate: Evidence for embryonic cardiac bradyarrhythmia due to inhibition of delayed rectifier K^+ channels resulting in hypoxia-reoxygenation damage. Teratology. 2001;**63**:152-160

[21] Nicolai J, Vles JS, Aldenkamp AP. Neurodevelopmental delay in children exposed to antiepileptic drugs in utero: A critical review directed at structural study-bias. Journal of the Neurological Sciences. 2008;**271**(1-2):1-14

[22] Tomson T, Battino D, Bonizzoni E, et al. Comparative risk of major congenital malformations with eight different antiepileptic drugs: A prospective cohort study of the EURAP registry. Lancet Neurology. 2018;**17**:530-538

[23] Kinney MO, Morrow J, Patterson CC, et al. Changing antiepilepsy drug prescribing trends in women with epilepsy in the UK and Ireland and the impact on major congenital malformations. Journal of Neurology, Neurosurgery, and Psychiatry. 2018;**89**:1320-1323

[24] Smith J, Whitehall J. Sodium valproate and the fetus: A case study and review of the literature. Journal of Neonatal Nursing. 2009;**28**:363-367

[25] Tomson T, Battino D, Bonizzoni E, Craig J, Lindhout D, Sabers A, et al. Dose-dependent risk of malformations with antiepileptic drugs: An analysis of data from the EURAP epilepsy and pregnancy registry. Lancet Neurology. 2011;**10**:609-617

[26] Crudup JB, Hartley BI, Keel BR, et al. Recognizing and treating valproic acid toxicity: A case report. Journal of Medical Cases. 2011;**5**:185-187

[27] Ibarra M, Vazquez M, Faqiolino P, Derendorf H. Sex related differences on valproic acid pharmacokinetics after oral single dose. Journal of Pharmacokinetics and Pharmacodynamics. 2013;**40**:479-486

[28] Jentink J, Loane MA, Dolk H, et al. Valproic acid monotherapy in pregnancy and major congenital malformations. The New England Journal of Medicine. 2010a;**362**:2185-2193

[29] Vajda FJE, O'Brien TJ, Lander CM, Graham J, Eadie MJ. Antiepileptic drug combinations not involving valproate and the risk of fetal malformations. Epilepsia. 2016;**57**(7):1048-1052

[30] Vossler DG. Comparative risk of major congenital malformations with 8 different antiepileptic drugs: A prospective cohort study of the EURAP registry. Epilepsy Currents. 2019;**19**(2):83-85. DOI: 10.1177/1535759719835353

[31] Tung EW, Winn LM. Epigenetic modifications in valproic acid-induced teratogenesis. Toxicology and Applied Pharmacology. 2010;**248**:201-209

[32] Hallene KL, Oby E, Lee BJ, et al. Prenatal exposure to thalidomide, altered vasculogenesis, and CNS malformations. Neurological Sciences. 2006;**142**:267-283

[33] Lenz W. Thalidomide and congenital abnormalities. Lancet. 1962;**1**:271-272

[34] Lenz W. A short history of thalidomide embryopathy. Teratology. 1988;**28**:203-215

[35] Miller MT, Stromland K. Teratogen update: Thalidomide: A review, with a focus on ocular findings and new potential uses. Teratology. 1999;**60**:306-321

[36] Vargesson N. Thalidomide-induced teratogenesis: History and mechanisms. Birth Defects Research. Part C, Embryo Today. 2015;**105**(2):140-156. DOI: 10.1002/bdrc.21096

[37] Ito T, Ando H, Handa H. Teratogenic effects of thalidomide: Molecular mechanisms. Cellular and Molecular Life Sciences. 2011;**68**:1569-1579

[38] D'Amato RJ, Loughnan MS, Flynn E, et al. Thalidomide is an inhibitor of angiogenesis. Proceedings of the National Academy of Sciences of the United States of America. 1994;**91**:4082-4085

[39] Vargesson N. Thalidomide-induced limb defects: Resolving a 50-year-old puzzle. BioEssays. 2009;**31**:1327-1336

[40] Yabu T, Tomimoto H, Taguchi Y, et al. Thalidomide-induced antiangiogenic action is mediated by ceramide through depletion of VEGF receptors, and is antagonized by sphingosine-1-phosphate. Blood. 2005;**106**:125-134

[41] Ujházy E, Mach M, Navarová J, Brucknerová I, Dubovický M. Teratology—Past, present and future. Interdisciplinary Toxicology. 2012;**5**(4):163-168

[42] De Santis M, Straface G, Cavaliere AF, Caruso A, Cichocki F, Venga L, et al. First trimester maternal thyroid X-ray exposure and neonatal birth weight. Reproductive Toxicology. 2005;**20**:3-4

[43] Goldberg-Stein S, Liu B, Hahn PF, Lee SI. Body CT during pregnancy: Utilization trends, examination indications, and fetal radiation doses. American Journal of Roentgenology. 2011;**196**:146-151

[44] Lazarus E, Debenedectis C, North D, Spencer PK, Mayo-Smith WW. Utilization of imaging in pregnant patients: 10-year review of 5270 examinations in 3285 patients—1997-2006. Radiology. 2009;**251**:517-524

[45] Brent RL. Utilization of developmental basic science principles in the evaluation of reproductive risks from pre- and postconception environmental radiation exposures. Teratology. 1999;**59**(4):182-204

[46] Patel SJ, Reede DL, Katz DS, Subramaniam R, Amorosa JK. Imaging the pregnant patient for nonobstetric conditions: Algorithms and radiation dose considerations. Radiographics. 2007;**27**(6):1705-1722

[47] Wrixon AD. New ICRP recommendations. Journal of Radiological Protection. 2008;**28**:161-168

[48] Ratnapalan S, Bona N, Chandra K, Koren G. Physicians' perceptions of teratogenic risk associated with radiography and CT during early pregnancy. American Journal of Roentgenology. 2004;**182**(5):1107-1109

[49] Nuyttens JJ, Prado KL, Jenrette JM, Williams TE. Fetal dose during radiotherapy: Clinical

implementation and review of the literature. Cancer Radiothérapie. 2002;**6**(6):352-357

[50] Streffer C, Shore R, Konermann G, Meadows A, Uma Devi P, Preston Withers J, et al. Biological effects after prenatal irradiation (embryo and fetus). A report of the international commission on radiological protection. Annals of the ICRP. 2003;**33**(1-2):5-206

[51] Kesmodel US et al. Are low-to-moderate average alcohol consumption and isolated episodes of binge drinking in early pregnancy associated with facial features related to fetal alcohol syndrome in 5-year-old children? Alcoholism, Clinical and Experimental Research. 2019;**43**(6):1199-1212. DOI: 10.1111/acer.14047

[52] Longhurst W, Ernst J, Burd L. Fetal alcohol exposure and development of the integument. Research and Reports in Neonatology. 2016;**6**:25-32. DOI: 10.2147/RRN.S99991

[53] Frias JL, Gilbert-Barness E. Human teratogens: Current controversies. Advances in Pediatrics. 2008;**55**:171-211

[54] Jańczewska I, Wierzba J, Cichoń-Kotek M, Jańczewska A. Fetal alcohol spectrum disorders—Diagnostic difficulties in the neonatal period and new diagnostic approaches. Developmental Period Medicine. 2019;**23**(1):60-66

[55] Pei J, Baugh L, Andrew G, Rasmussen C. Intervention recommendations and subsequent access to services following clinical assessment for fetal alcohol spectrum disorders. Research in Developmental Disabilities. 2017;**60**:176-186. DOI: 10.1016/j.ridd.2016.11.007

[56] Streissguth AP, Bookstein FL, Barr HM, Sampson PD, O'Malley K, Young JK. Risk factors for adverse life outcomes in fetal alcohol syndrome and fetal alcohol effects. Journal of Developmental and Behavioral Pediatrics. 2004;**25**(4):228-238. DOI: 10.1097/00004703-200408000-00002

[57] Popova S, Lange S, Shield K, Mihic A, Chudley AE, Mukherjee RA, et al. Comorbidity of fetal alcohol spectrum disorder: A systematic review and meta-analysis. Lancet. 2016;**387**(10022):978-987. DOI: 10.1016/s0140-6736(15)01345-8

[58] Shi M, Wehby GL, Murray JC. Review on genetic variants and maternal smoking in the etiology of oral clefts and other birth defects. Birth Defects Research. Part C, Embryo Today. 2008;**84**(1):16-29

[59] Holitzki H, Dowsett LE, Spackman E, Noseworthy T, Clement F. Health effects of exposure to second- and third-hand marijuana smoke: A systematic review. CMAJ Open. 2017;**5**(4):E814-E822

[60] Lee LJ, Lupo PJ. Maternal smoking during pregnancy and the risk of congenital heart defects in offspring: A systematic review and metaanalysis. Pediatric Cardiology. 2013;**34**(2):398-407

[61] Sabbagh HJ, Hassan MH, Innes NP, Elkodary HM, Little J, Mossey PA. Passive smoking in the etiology of non-syndromic orofacial clefts: A systematic review and meta-analysis. PLoS One. 2015;**10**(3):e0116963

[62] Martelli DR, Coletta RD, Oliveira EA, et al. Association between maternal smoking, gender, and cleft lip and palate. Brazilian Journal of Otorhinolaryngology. 2015;**81**(5):514-519

[63] Hackshaw A, Rodeck C, Boniface S. Maternal smoking in pregnancy and birth defects: A systematic review based on 173687

malformed cases and 11.7 million controls. Human Reproduction Update. 2011;**17**:589-604

[64] Pei L, Kang Y, Cheng Y, Yan H. The association of maternal lifestyle with birth defects in Shaanxi Province, Northwest China. PLoS One. 2015;**10**(9):e0139452

[65] Pankhurst MW, Gell DA, Butler CW, Kirkcaldie MT, West AK, Chung RS. Metallothionein (MT)-I and MT-II expression are induced and cause zinc sequestration in the liver after brain injury. PLoS One. 2012;7(2):e31185. DOI: 10.1371/journal. pone.0031185

[66] Kimura T, Itoh N. Function of metallothionein in gene expression and signal transduction: Newly found protective role of metallothionein. Journal of Health Science. 2008:**54**:251-260

[67] Hanna LA, Clegg MS, Momma TY, Daston GP, Rogers JM, Keen CL. Zinc influences the in vitro development of peri-implantation mouse embryos. Birth Defects Research. Part A, Clinical and Molecular Teratology. 2003;**67**:414-420

[68] Foster M, Samman S. Zinc and redox signaling: Perturbations associated with cardiovascular disease and diabetes mellitus. Antioxidants & Redox Signaling. 2010;**13**(10):1549-1573

[69] Adamo AM, Zago MP, Mackenzie GG, Aimo L, Keen CL, Keenan A, et al. The role of zinc in the modulation of neuronal proliferation and apoptosis. Neurotoxicity Research. 2010;**17**:1-14

[70] Clegg MS, Hanna LA, Niles BJ, Momma TY, Keen CL. Zinc deficiency-induced cell death. IUBMB Life. 2005;**57**:661-669

[71] Hess SY, King JC. Effects of maternal zinc supplementation on pregnancy and lactation outcomes. Food and Nutrition Bulletin. 2009;**30**:S60-S78

[72] Shah D, Sachdev HP. Zinc deficiency in pregnancy and fetal outcome. Nutrition Reviews. 2006;**64**:15-30

[73] Neville AJ, de Walle HEK. Prevention of neural tube defects by periconceptional folic acid supplementation in Europe. Gynaecology Forum. 2010;**15**:16-20

[74] Kocylowski R, Grzesiak M, Gaj Z, Lorenc W, Bakinowska E, Barałkiewicz D, et al. Associations between the level of trace elements and minerals and folate in maternal serum and amniotic fluid and congenital abnormalities. Nutrients. 2019;**11**(2). pii: E328. DOI: 10.3390/nu11020328

[75] Bitzer J, von Stenglin A, Bannemerschultz R. Women's awareness and periconceptional use of folic acid: Data from a large European survey. International Journal of Women's Health. 2013;**5**:201-213

[76] Goyette D, Summer JS, Milos R. Human methylenetetrahydrofolate reductase: Isolation of cDNA, mapping and mutation identification. Nature Genetics. 1994;**7**:195-200

[77] De Wals P, Tairou F, van Allen M, Uh SH, Lowry RB, Sibbald B, et al. Reduction in neural-tube defects after folic acid fortification in Canada. The New England Journal of Medicine. 2007;**357**:135-142

[78] Wang H, De Steur H, Chen G, Zhang X, Pei L, Gellynck X, et al. Effectiveness of folic acid fortified flour for prevention of neural tube defects in high risk region. Nutrients. 2016;**8**(3):1-11

[79] Yan J, Zheng YZ, Cao LJ, Liu YY, Li W, Huang GW. Periconceptional folic acid supplementation in Chinese women: A cross-sectional study.

Biomedical and Environmental Sciences. 2017;**30**(10):737-748

[80] Wang H, Ge X, Zhu B, Xuan Y, Huang K, Rutayisire E, et al. Maternal continuing folic acid supplementation after the first trimester of pregnancy ıncreased the risk of large-for gestational-age birth: A population-based birth cohort study. Nutrients. 2016;**8**(8):1-11

[81] Baohong M, Jie Q, Nan Z, Yawen S, Wei D, Xiaochun H, et al. Maternal folic acid supplementation and dietary folate intake and congenital heart defects. PLoS One. 2017;**12**(11):1-14

[82] Morgan SC, Relaix F, Sandell LL, Loeken MR. Oxidative stress during diabetic pregnancy disrupts cardiac neural crest migration and causes outflow tract defects. Birth Defects Research. Part A, Clinical and Molecular Teratology. 2008;**82**:453-463

[83] Ornoy A, Reece EA, Pavlinkova G, et al. Effect of maternal diabetes on the embryo, fetus, and children: Congenital anomalies, genetic and epigenetic changes and developmental outcomes. Birth Defects Research. Part C, Embryo Today. 2015;**105**:53-72

[84] Kappen C, Salbaum JM. Gene expression in teratogenic exposures: A new approach to understanding individual risk. Reproductive Toxicology. 2014;**45**:94-104

[85] Kappen C, Kruger C, MacGowan J, Salbaum JM. Maternal diet modulates the risk for neural tube defects in a mouse model of diabetic pregnancy. Reproductive Toxicology. 2011;**31**:41-49

[86] Christianson AHC, Modell B. March of Dimes Global Report on Birth Defects: The Hidden Toll of Dying and Disabled Children. NY: White Plains; 2006. p. 76

[87] Ornoy A, Reece EA, Pavlinkova G, Kappen C, Miller RK. Effect of maternal diabetes on the embryo, fetus, and children: Congenital anomalies, genetic and epigenetic changes and developmental outcomes. Birth Defects Research Part C: Embryo Today: Reviews. 2015;**105**(1):53-52. DOI: 10.1002/bdrc.21090

[88] Prick BW, Hop WC, Duvekot JJ. Maternal phenylketonuria and hyperphenylalaninemia in pregnancy: Pregnancy complications and neonatal sequelae in untreated and treated pregnancies. The American Journal of Clinical Nutrition. 2012;**95**:374-382

[89] Cetin I, Corbetta C, Sereni LP, Marconi AM, Bozzetti P, Pardi G, et al. Umbilical amino acid concentration in normal and growth retarded fetuses sampled in utero by cordocentesis. American Journal of Obstetrics and Gynecology. 1990;**162**:253-261

[90] Waisbren SE, Rohr F, Anastasoaie V, Brown M, Harris D, Ozonoff A, et al. Maternal phenylketonuria: Long-term outcomes in offspring and post-pregnancy maternal characteristics. JIMD Reports. 2015;**21**:23-33

[91] Koch R, Azen C, Friedman E, Hanley W, Levy H, Matalon R, et al. Research design, organization, and sample characteristics of the maternal PKU collaborative study. Pediatrics. 2003;**112**:1519-1522

[92] Pereira L. Congenital viral infection: Traversing the uterine-placental interface. Annual Review of Virology. 2018;**5**(1):273-299

[93] Neu N, Duchon J, Zachariah P. TORCH infections. Clinics in Perinatology. 2015;**42**(1):77-103. viii. DOI: 10.1016/j.clp.2014.11.001

[94] Chen L, Liu J, Shi L, Song Y, Song Y, Gao Y, et al. Seasonal influence on TORCH infection and analysis of

multi-positive samples with indirect immunofluorescence assay. Journal of Clinical Laboratory Analysis. 2019;**33**(4):e22828. DOI: 10.1002/jcla.22828

[95] Robert-Gangneux F, Darde ML. Epidemiology of and diagnostic strategies for toxoplasmosis. Clinical Microbiology Reviews. 2012;**25**(2):264-296. DOI: 10.1128/CMR/05013-11

[96] Oz HS. Maternal and congenital toxoplasmosis, currently available and novel therapies in horizon. Frontiers in Microbiology. 2014;**5**:385. DOI: 10.3389/fmicb.2014.00385

[97] Hökelek M. In: Bronze MS, ed. Toxoplasmosis Clinical Presentation. 2017. Available from: https://emedicine.medscape.com/article/229969-clinical [Accessed: 19 September 2017]

[98] Torgerson PR, Mastroiacovo P. The global burden of congenital toxoplasmosis: A systematic review. Bulletin of the World Health Organization. 2013;**91**(7):501-508. DOI: 10.2471/BLT.12.111732

[99] Kwofie KD, Ghansah A, Osei JH, Frempong KK, Obed S, Frimpong EH, et al. Indication of risk of mother-to-child *Toxoplasma gondii* transmission in the Greater Accra region of Ghana. Maternal and Child Health Journal. 2016;**20**(12):2581-2588

[100] Wallon M, Garweg JG, Abrahamowicz M, Cornu C, Vinault S, Quantin C, et al. Ophthalmic outcomes of congenital toxoplasmosis followed until adolescence. Pediatrics. 2014;**133**:e601-e608. DOI: 10.1542/peds.2013-2153

[101] El Bissati K, Levigne P, Lykins J, et al. Global initiative for congenital toxoplasmosis: An observational and international comparative clinical analysis. Emerging Microbes and Infections. 2018;**7**(1):165. DOI: 10.1038/s41426-018-0164-4 [Published: 27 September 2018]

[102] McLeod R, Lykins J, Noble AG, Rabiah P, Swisher CN, Heydemann PT, et al. Management of congenital toxoplasmosis. Current Pediatrics Reports. 2014;**2**:166-194. DOI: 10.1007/s40124-014-0055-7

[103] McLeod R, Kieffer F, Sautter M, Hosten T, Pelloux H. Why prevent, diagnose and treat congenital toxoplasmosis? Memórias do Instituto Oswaldo Cruz. 2009;**104**(2):320-344

[104] Maldonado YA, Read JS. Diagnosis, treatment, and prevention of congenital toxoplasmosis in the United States committee on infectious diseases. The Journal of Pediatrics. 2017;**139**(2):e20163860. DOI: 10.1542/peds.2016-3860

[105] George S, Viswanathan R, Sapkal G. Molecular aspects of the teratogenesis of rubella virus. Biological Research. 2019;**52**. DOI: 10.1186/s40659-019-0254-3

[106] Beatch MD, Everitt JC, Law LJ, Hobman TC. Interactions between Rubella virus capsid and host protein p32 are important for virus replication. Journal of Virology. 2005;**79**(16):10807-10820. DOI: 10.1128/JVI.79.16.10807-10820.2005

[107] Shukla S, Maraqa NF. Congenital Rubella. NCBI Bookshelf. A Service of the National Library of Medicine, National Institutes of Health; 2018

[108] Crane J, Mundle W, Boucoiran I, Maternal Fetal Medicine Committee. Parvovirus B19 infection in pregnancy. Journal of Obstetrics and Gynaecology Canada. 2014;**36**:1107-1116

[109] Hellmund A, Geipel A, Berg C, Bald R, Gembruch U. Early intrauterine transfusion in fetuses with severe anemia caused by parvovirus B19

infection. Fetal Diagnosis and Therapy. 2018;**43**:129-137

[110] Gilarranz R, Chamizo F, Hernandez-Febles M, Valle L, Pena-Lopez MJ. Parvovirus B19 congenital infection. Infectious Diseases (London). 2016;**48**: 566-568

[111] Mackie FL, Pretlove SJ, Martin WL, Donovan V, Kilby MD. Fetal intracardiac transfusions in hydropic fetuses with severe anemia. Fetal Diagnosis and Therapy. 2015;**38**:61-64

[112] Dollard SC, Grosse SD, Ross DS. New estimates of the prevalence of neurological and sensory sequelae and mortality associated with congenital cytomegalovirus infection. Reviews in Medical Virology. 2007;**17**:355-363

[113] Marsico C, Kimberlin DW. Congenital cytomegalovirus infection: Advances and challenges in diagnosis, prevention and treatment. Italian Journal of Pediatrics. 2017;**43**(1):38

[114] Gabrielli L, Losi L, Varani S, Lazzarotto T, Eusebi V, Landini MP. Complete replication of human cytomegalovirus in explants of first trimester human placenta. Journal of Medical Virology. 2001;**64**:499-504

[115] Gabrielli L, Lazzarotto T, Foschini M, Lanari M, Guerra B, Eusebi V, et al. Horizontal in utero acquisition of cytomegalovirus infection in a twin pregnancy. Journal of Clinical Microbiology. 2003;**41**:1329-1331. DOI: 10.1128/JCM.41.3.1329-1331.2003

[116] Lanzieri TM, Dollard SC, Bialek SR, Grosse SD. Systematic review of the birth prevalence of congenital cytomegalovirus infection in developing countries. International Journal of Infectious Diseases. 2014;**22**:44-48

[117] Kimberlin DW, Jester PM, Sanchez PJ, et al. Valganciclovir for symptomatic congenital cytomegalovirus disease. The New England Journal of Medicine. 2015;**372**:933-943

[118] Revello MG, Tibaldi C, Masuelli G, et al. Prevention of primary cytomegalovirus infection in pregnancy. eBioMedicine. 2015;**2**:1205-1210

[119] Blumental S, Lepage P. Management of varicella in neonates and infants. BMJ Paediatrics Open. 2019;**3**:e000433. DOI: 10.1136/bmjpo-2019-000433

[120] Kido S, Ozaki T, Asada H, Higashi K, Kondo K, Hayakawa Y, et al. Detection of varicella-zoster virus (VZV) DNA in clinical samples from patients with VZV by the polymerase chain reaction. Journal of Clinical Microbiology. 1991;**29**(1):76-79

[121] Ahn KH, Park Y-J, Hong S-C, Lee EH, Lee J-S, Oh M-J, et al. Congenital varicella syndrome: A systematic review. Journal of Obstetrics and Gynaecology. 2016;**36**(5):563-566. DOI: 10.3109/01443615.2015.1127905

[122] Cohen A, Moschopoulos P, Stiehm RE, Koren G. Congenital varicella syndrome: The evidence for secondary prevention with varicella-zoster immune globulin. Canadian Medical Association Journal. 2011;**183**:204-208

[123] Pomar L, Malinger G, Benoist G, Carles G, Ville Y, Rousset D, et al. Association between Zika virus and fetopathy: A prospective cohort study in French Guiana. Ultrasound in Obstetrics & Gynecology. 2017;**49**(6):729-736. DOI: 10.1002/uog.17404. Epub 2017

Chapter 3

The Impact of Maternal Infection on the Neonate

Hellen Ko, Walter Dehority and Jessie R. Maxwell

Abstract

Maternal infection is a common occurrence during pregnancy, with a substantial impact on the infant. Some infections result in impaired development *in utero* and even death of the fetus. Other infections may be insidious in the mother but result in growth impairment and hearing loss in the infant. A growing body of evidence suggests that even infections such as chorioamnionitis, thought to have no long-term impact on the infant, may alter fetal development. This chapter will review congenital infections and their impact on neonatal outcomes, as well as newer findings suggesting that acute infection may result in adverse changes in the infant. We will explore novel mechanisms of pathogenesis and virulence, as well as areas that continue with ongoing research.

Keywords: pregnancy, infection, neurodevelopment, chorioamnionitis, TORCH infections, Zika

1. Introduction

Maternal infections during pregnancy can have a direct impact on the developing fetus and in some infections can result in fetal demise. It is extremely important to screen women for infections when it is available and practical and to treat when necessary. The current screening tests recommended by the American College of Obstetricians and Gynecologists include rubella, hepatitis B, hepatitis C, human immunodeficiency virus (HIV), Group B streptococcus (GBS), tuberculosis and sexually transmitted infections including syphilis, chlamydia, and gonorrhea if risk factors are present [1]. The incidence of congenital infections in infants varies, with syphilis increasing dramatically from 639 cases in 2016 to over 1300 cases in 2018 in the United States [2]. Additionally, congenital cytomegalovirus, varicella zoster virus and herpes simplex virus diagnoses have increased over the last five decades [3]. Rubella has decreased since the introduction of Rubella immunization; prior to utilization of the immunization, over 100,000 infants were born worldwide with congenital rubella syndrome (CRS). By 2014, a 95% decrease in cases of CRS was observed in countries that followed the immunization schedule [4]. Thus, it is critically important that research efforts continue to prioritize the development of immunizations and treatments plans for all viruses that can result in congenital fetal infection in an attempt to minimize the substantial long-term morbidities that result.

IntechOpen

2. Chorioamnionitis/intra-amniotic infection (IAI)

Chorioamnionitis is the term that has been used for decades to describe infection and/or inflammation of the chorion, amnion, or both. This has been further delineated into a "clinical" diagnosis based on maternal symptoms, and a "histological" diagnosis based on the pathology of the placenta following delivery. Clinical signs and symptoms are used to diagnose clinical chorioamnionitis, and include maternal fever, uterine fundal tenderness, maternal and/or fetal tachycardia and purulent amniotic fluid [5]. The most common bacterial organisms to cause chorioamnionitis are *Ureaplasma urealyticum* and *Mycoplasma hominis*. Histological chorioamnionitis is diagnosed by observing neutrophil infiltration into the chorion and amnion [6]. The variation in the definition of chorioamnionitis has resulted in confusion in neonatal management as well as difficulty in assessing the long-term impact of chorioamnionitis on development. Therefore, intra-amniotic infection (IAI) has been developed to replace the prior diagnosis of chorioamnionitis [7].

IAI was updated in 2017 by the American College of Obstetricians and Gynecologists into three categories which are readily diagnosed. Isolated maternal fever (IMF) is the first category, in which the mother has a single intrapartum temperature of ≥39.0°C or a temperature of 38.0–38.9°C that persists for 30 min, with treatment recommendations including the consideration of broad-spectrum antibiotics [7, 8]. Given the numerous potential causes of maternal fever, the utilization of antibiotics is at the providers' discretion. Suspected IAI is diagnosed when the mother has an elevated temperature (≥39.0°C) or a slightly elevated temperature (38.0–38.9°C) along with one of the following risk factors: maternal leukocytosis, purulent cervical drainage or fetal tachycardia [7, 8]. Confirmed IAI is diagnosed with a positive amniotic fluid test or placental pathology demonstrating histologic evidence of infection [7]. Similar to the previously used histological chorioamnionitis, a criticism of this diagnosis is that it is made after the clinical situation has resolved, and thus does not aid in the acute management of the mother or the infant. Both suspected and confirmed IAI diagnoses should result in treatment with intrapartum antibiotics and antipyretics [7].

IAI is present in nearly 50% of very early preterm birth [9], after which multiple complications can occur and a wide array of neonatal morbidities and mortalities are observed. This has led to speculation that IAI is directly impacting the fetal and neonatal development and outcomes, as well as potentially resulting in preterm birth, which then impacts development and outcomes. The majority of studies that have investigated this question utilized diagnoses of chorioamnionitis, which included both clinical and histological cases. Given the variation of diagnoses included in these studies, it is not surprising that the results have also been varied. A large study of 2390 extremely preterm infants (born <27 weeks' gestational age) from sixteen centers across the United States found infants exposed to histological and clinical chorioamnionitis had an increased risk of cognitive impairment at 18–22 months' corrected age [10]. A separate study of 350 infants found that while gestational age was significantly lower among those with exposure to histological chorioamnionitis, there was no association with intraventricular hemorrhage, white matter injury around birth, or differences in cognitive or motor outcomes at 18–24 months' corrected age [11]. Additional studies have found weak causal or associative roles of chorioamnionitis with cerebral palsy risk [12] and no increased risk of white matter injury on magnetic resonance imaging (MRI) following histological chorioamnionitis in premature infants [13]. Additional investigation is required with the new IAI definitions to determine if there are consistent findings with developmental outcomes in those diagnosed with IAI.

3. TORCH infections

TORCH infection is a mnemonic that has classically been used to describe congenital infections that can impact fetal development. In the past, TORCH represented Toxoplasmosis, Other (syphilis, varicella-zoster, parvovirus B19 and newer pathogens such as Zika), Rubella, Cytomegalovirus and Herpes Simplex Virus. However, as more pathogens are being discovered and the "other" category is expanding, some experts feel the mnemonic is not as relevant today.

3.1 Toxoplasmosis

Toxoplasma gondii is an obligate intracellular protozoan which typically causes mild illness in most immunocompetent individuals [14, 15]. While a large portion of infected children and adults are asymptomatic, Toxoplasmosis is considered one of the major causes of death linked to foodborne illness in the United States. If an immunocompromised individual, pregnant woman, or fetus/infant acquires the infection, there can be severe, even fatal, consequences [14, 15]. Illness can range from non-specific systemic symptoms such as fever, lymphadenopathy and hepato-splenomegaly to congenital toxoplasmosis (CT), which is classically described as a triad of chorioretinitis, intracranial calcifications and hydrocephalus. CT can lead to loss of vision and hearing, decreased cognitive function, and neurodevelopmental delay if untreated [14, 16–18].

T. gondii exists in three forms: tachyzoite, bradyzoite, and sporozoite. The definitive hosts are members of the *Felidae* family, but warm-blood mammals can also serve as intermediate hosts [17]. Felines can acquire *T. gondii* through the ingestion of tissue cysts containing bradyzoites in infected prey or through the ingestion of oocysts containing sporozoites in anything contaminated with feces from an infected cat. They can excrete un-sporulated oocysts in their stools 3–30 days after infection and can shed for 7–14 days. If in the right climate (such as warm and humid), the oocysts can sporulate for 1–5 days, after which they can remain infectious for years. If the tissue cysts found in intermediate hosts or the sporulated oocysts are ingested by humans, they transform into active tachyzoites. The tachyzoites then primarily infect the central nervous system, eyes, musculo-skeletal system, and placenta by infecting nucleated host cells to bypass the blood brain barrier and placental barricade. Incubation is 7 days with a range of 4–21 days [14, 15, 18].

For pregnant women who have an acute infection with *T. gondii*, the timing can be crucial and dictates the treatment course. Typically, the earlier in pregnancy that acute infection occurs, the lower the rate of transmission to the fetus. Unfortunately, there is an increased severity of illness if transmission occurs earlier in the pregnancy [14, 15]. The reverse is true for infection later in pregnancy (such as during the third trimester), during which there is a high rate of transmission but with less severe illness in the fetus.

The diagnosis of primary or latent infection is made primarily using serologic tests. Toxoplasma-specific Immunoglobulin G (IgG) and Immunoglobulin M (IgM) can be performed routinely at non-reference laboratories. Any positive IgM results are then submitted to reference laboratories that can perform additional testing for confirmation [18]. If a pregnant woman is found to have acute infection, then an amniocentesis can be performed, and the fluid can be sent for polymerase chain reaction (PCR) testing. If the PCR is negative and the fetus is believed to have not acquired the infection, the next best step is treatment in the mother with spiramycin in an attempt to prevent transmission [14, 15, 17, 18]. If, however, the fetus is thought to be infected, then the mother is started on a combination of pyrimethamine, sulfadiazine, and folinic acid. Spiramycin, a primarily bacteriostatic macrolide that has activity against some gram-negative and gram-positive organisms as well as some spirochetes, is unable to cross the placenta whereas the combination of anti-parasitic medications can cross the placenta and thus can aide in treatment of the fetus [18, 19]. The combination is also used for fetal infection confirmed at or after 18 weeks of gestation or maternal infection acquired during the third trimester [14, 17, 18]. As untreated CT can lead to fetal demise or death within the first few days of life, and chorioretinitis can develop in a significant proportion of infants whose mothers were untreated, it is imperative to diagnose and start treatment in a timely manner [18].

Once an infant with suspected CT is born, he or she should be thoroughly examined and evaluated. Serologies, a complete blood count (CBC), hepatic function tests, blood PCR, urine PCR, cerebrospinal fluid (CSF) PCR, and CSF studies including glucose level, protein, and cell count, should be sent [18]. The newborn should also have ophthalmologic, auditory, and neurologic evaluation including imaging of the brain [18]. Infected infants should receive treatment regardless of any clinically apparent symptoms, as a large proportion of infants with asymptomatic CT at birth go on to develop visual/hearing impairment, learning disabilities, and psychomotor delay [15, 16, 18, 20]. Treatment consists of the same anti-parasitic combination of pyrimethamine, sulfadiazine, and folinic acid [18, 19]. If CSF studies show an elevated protein concentration (greater than 1 g/dL) or there is evidence of severe chorioretinitis, then a corticosteroid such as prednisone is added until there is a decrease in protein concentration in the CSF or resolution of severe chorioretinitis [15, 18, 19]. Treatment is continued at least though 12 months of age, with consideration of shorter treatment duration for infants who remain asymptomatic for the first three months of life [18, 19]. For those infants who are asymptomatic with positive Toxo-specific IgG but negative IgM and Immunoglobulin A (IgA), there should be repeat IgG testing every four to six weeks until disappearance of IgG. There is no clear consensus on the treatment of these infants [18, 19].

Studies looking at the outcomes of infants with CT have shown significantly better neurologic and developmental outcomes in those that were treated than those who were not [21]. It is important to note that compared to their uninfected siblings, the children that received treatment had a lower level of cognitive function though there was no deterioration over time. In terms of ophthalmologic outcomes, it was found that when followed up to 22 years of age, new ocular lesions could be detected in adolescence which points to the importance of continued ophthalmologic evaluation.

3.2 Other: syphilis, varicella-zoster, parvovirus B19, Zika

3.2.1 Syphilis

Treponema pallidum, a thin, motile spirochete, is the organism that causes syphilis [18], a sexually transmitted infection that can also result in congenital infection to a fetus. While there was initially a decline in the cases of syphilis observed in the United States in 2000–2001, an alarming resurgence has recently been noted. There has been an increase of 72% in the number of reported primary and secondary cases in the United States from 2013 to 2017, with the number of congenital syphilis cases increasing more than 150% from 2013 to 2018 [22–24]. It is thought that the increase in methamphetamine use, having sex with a person who injects drugs, injection drug use and heroin use are the primary factors that are leading to this dramatic increase in syphilis cases [22, 25].

Acquired syphilis is typically divided into three stages: primary, secondary and latent. During the primary stage, painless indurated ulcers form on the skin or mucous membranes of the areas exposed and heal spontaneously in a few weeks. The secondary stage, typically 1–2 months after the primary stage, is characterized by a maculopapular rash that typically includes the palms and soles, lymphadenopathy and mucocutaneous lesions including condylomata lata [18]. Finally, the latent stage occurs when there are no clinical signs or symptoms of infection, but an individual remains seroreactive [18]. *T. pallidum* can infect the central nervous system (CNS) during any stage, resulting in neurosyphilis. Transmission to the fetus during pregnancy can occur at any point, with primary and secondary syphilis having the highest rates of transmission at 60–100% [18].

It is recommended that all women be screened for syphilis early in pregnancy with a nontreponemal test, with repeat testing later in pregnancy for high risk individuals. These tests include the Venereal Disease Research Laboratory (VDRL) slide test and the rapid plasma reagin (RPR) test [18]. These nontreponemal tests utilize an antigen that reacts in the presence of antibodies (to syphilis). However, given that the antigen is not specific for syphilis and is a component of cell membranes, false positives may result from other infections including varicella and measles, or by tissue damage observed in connective tissue disease and even pregnancy itself [26]. Therefore, a positive nontreponemal test should be followed by a confirmatory test such as fluorescent treponemal antibody absorption (FTA-ABS) or *T. pallidum* particle agglutination (TP-PA) tests. Additionally, any person found positive for syphilis based on screening and confirmatory testing should also be screened for human immunodeficiency virus (HIV) given the high rate of co-infection.

Treatment for syphilis is parenteral penicillin G; if an individual is allergic to penicillin G, they should undergo desensitization due to the lack of proven efficacy of alternative agents in this setting. Lack of treatment during pregnancy can result in stillbirth and neonatal death in nearly 40% of women with primary and secondary stage disease, 40% of infants being infected and only 20% of infants being healthy and uninfected [27]. Additionally, fetal infection can result in anemia, hepatomegaly and hydrops [24]. Treatment of the infant should not be delayed, as early treatment may prevent neurologic sequelae [24].

A serological diagnosis is made on the infant if the nontreponemal titer (VDRL or RPR) is fourfold higher than that of the mother (both samples should be obtained around the same time), if the nontreponemal titer persists or increases after birth, or if the treponemal antibody titer (FTA-ABS or TP-PA) remains positive at 12–18 months of age. The choice of test on the infant is dependent on the test that the mother had received, as the titers will need to be compared [18]. A complete evaluation, including complete blood cell count (CBC), liver function tests, obtaining cerebrospinal fluid (CSF) to test for VDRL reactivity, ophthalmologic examination and long-bone radiographs to assess abnormal ossification, radiolucencies or dislocation of epiphyses is then needed [28]. Neuroimaging should be considered if there are any concerns for central nervous system involvement [18]. Ten days of treatment with parenteral penicillin G is typically used in infected infants, with close follow up required. Titers should be repeated by 3 months of age and noted to be declining, with nonreactivity noted by 6 months of age [28]. If the mother received appropriate treatment that was administered >4 weeks before delivery, and the infant has a normal physical examination with the titer equal to or less than fourfold the maternal titer, then no evaluation is recommended. However, inadequate treatment in the mother should result in evaluation of the infant and treatment with penicillin G for 10 days [28].

Clinically, nearly half of infants do not have any apparent signs of infection, although bone lesions and hematologic and hepatobiliary abnormalities may be present, with hepatomegaly one of the most common findings [24, 29]. Infants that develop symptoms may have rhinitis in the first week of life, in which persistent white discharge ("snuffles") occurs which contains spirochetes [29]. Additional symptoms can include generalized lymphadenopathy and a maculopapular rash [29]. Long term outcomes of infants not appropriately treated can include sensorineural hearing loss, interstitial keratitis, secondary glaucoma, corneal scarring, vision impairment, Hutchinson teeth (smaller teeth that are widely spaced with notches), saber shins (sharp anterior bowing of the tibia), frontal bossing, saddle nose, gummas (soft, non-cancerous growth) and scarring [29]. Life-long disabilities can occur in congenital syphilis infections if infants are not appropriately screened and treated [28].

3.2.2 Varicella-zoster

Varicella-zoster virus (VZV) is a herpesvirus that is transmitted by respiratory droplets, direct contact with skin lesions, and transplacentally during pregnancy [30]. Infants that are exposed to VZV during the last few weeks of pregnancy may develop neonatal varicella which can be quite severe; congenital varicella syndrome (CVS) develops in infants exposed during the pregnancy, with the risk being highest if the exposure occurs in the first trimester [30]. Infants exposed after 20 weeks' gestation only have about 2% chance of developing CVS [31]. Infants with CVS most commonly have skin lesions in a dermatomal distribution followed by neurologic defects, eye disease and skeletal anomalies [31]. Neurologic defects can include cerebral cortical atrophy and ventriculomegaly. Unfortunately, CVS is fatal in about 30% of cases within the first month of life [32].

The monovalent vaccine approved in 1995 and the quadrivalent vaccine introduced in 2005 have impacted the prevalence of congenital infection as seroprotection is nearly 100% after 2 doses of the vaccine [18]. Thus, at this time, CVS is considered an extremely rare disorder.

3.2.3 Parvovirus B19

Human parvovirus B19 is a nonenveloped, single-stranded deoxyribonucleic acid (DNA) virus with humans as the only host [18]. The virus replicates in erythrocyte precursors and is transmitted via respiratory tract secretions, exposure to blood or blood products, and vertically [18]. While it often causes a mild respiratory tract infection with a "slapped cheek" rash, it can be lethal to a fetus, with the risk of death being as high as 10% [33]. The incidence of parvovirus B19 infection during pregnancy is 3–4%, with the transplacental transmission rate approaching 30% [34]. Fortunately, approximately 50–75% of women of reproductive age are immune to parvovirus B19 [35]. The timing of infection during pregnancy does alter the risk of fetal death, with first trimester infections resulting in up to 71% risk of fetal loss [34]. The difficulty in diagnosing the virus during pregnancy arises in the lack of symptoms that most adults experience, and as many as 70% of women would have no symptoms if infected during pregnancy [34]. Arthropathies are one of the most common symptoms and should raise suspicion for possible infection [34]. Additionally, the presence of fetal ascites or pericardial effusions on ultrasound should trigger high suspicion as well [33].

Fetal hydrops, or abnormal accumulation of fluid/edema in two or more compartments, is common in the setting of Parvovirus B19 infection, with a meta-analysis finding a 9.3% pooled incidence, as well as an increased risk of fetal loss, spontaneous abortion and stillbirth [36]. Parvovirus B19 is among the most common causes of non-immune fetal hydrops, and while spontaneous resolution of infection can occur, only about 5% of cases with hydrops will show spontaneous resolution of the infection with disappearance of hydrops on follow up ultrasounds [37].

Severe anemia and thrombocytopenia occur *in utero* following parvovirus B19 infection, along with myocardial dysfunction [38]. These factors together are likely the etiology of the fetal hydrops. *In utero* transfusions (IUT) are often necessary and reduce mortality rates when compared to expectant management. A meta-analysis found IUT was performed in 78% of hydropic fetuses compared to 29% of non-hydropic fetuses, with the difference likely due to the hydropic fetuses at higher risk of demise [37]. Complications may occur in up to 5% of cases, especially if the fetus is likely more sensitive to vascular overload [38]. Thus, intrauterine exchange transfusions (IUET) have also been attempted in cases of fetal hydrops in the setting of parvovirus B19 infection. Unfortunately, thus far it results in similar survival rates as IUT and does not seem to be clinically superior as a treatment modality [38].

Longer-term testing reveal abnormal neurodevelopment following intrauterine parvovirus B19 infections in those also diagnosed with hydrops. Brain abnormalities including parenchymal calcifications, venous infarction, arterial infarction, cerebellar hemorrhage, and cortical malformations including diffuse cortical dysplasia and polymicrogyria have been described in congenital parvovirus infections [39]. If there are no abnormalities on imaging and hydrops resolves prior to delivery, one study found normal neurodevelopment in survivors at 1- and 5-year follow-up [40]. While the overall risk of mortality and morbidity are high, there is the potential for a normal outcome in select cases of congenital parvovirus infections.

3.2.4 Zika

Zika virus, ZIKV, is an emerging flavivirus that first became apparent internationally after Brazil declared a national public health emergency in 2016 followed by the World Health Organization declaring the outbreak a public health event of international concern [41]. The virus was first identified in 1947 in Uganda, after which cases of human infection have been infrequent and fairly localized [41]. ZIKV is transmitted by infected *Aedes* spp. mosquitoes, sexual contact and blood transfusions [42]. Around 80% of ZIKV that occur in adults are asymptomatic, with other cases having a mild febrile illness, headache, rash, fever and conjunctivitis [42]. However, severe neurologic sequalae can also occur in adults.

Congenital Zika syndrome (CZS) is variable in the presentation and severity with only a subset of infants that were exposed having apparent signs and symptoms at birth [41]. Infants exposed to ZIKV *in utero* are expected to survive, however a severe phenotype can result, particularly when exposure occurs in the first trimester [43]. ZIKV replication in brain tissue can continue after birth, and thus infants that are initially asymptomatic may develop symptoms within the first year of life [41]. The phenotype of CZS appears to consist of severe microcephaly and possibly a partially collapsed skull, thin cerebral cortices with subcortical calcifications, macular scarring, congenital contractures and marked early hypertonia [41]. Microcephaly is the most common symptom, occurring in up to 91% of CZS, and is often severe with the mean occipitofrontal head circumference falling 3–4 standard deviations below normal [43]. Both the central and peripheral nervous systems are impacted, with resultant effects on musculoskeletal, auditory and ophthalmologic systems and symptoms including conductive hip dysplasia, abnormal posturing of extremities, conductive hearing loss and abnormalities of the retina and optic nerve [43]. Up to 55% of infants with CZS have structural ocular abnormalities, making visual screening and interventions critically important to occur early in life to allow for neuroplasticity optimizing the outcomes [44]. This has led to the recommendation of any infant with suspected CZS or exposure to ZIKV to have an ocular examination before hospital discharge and again at 3 months of age [44].

A meta-analysis of 42 articles revealed the most common brain abnormalities following ZIKV exposure *in utero*, including decreased brain volume, increased extra-axial cerebrospinal fluid space, subcortical calcifications, microcephaly, ventriculomegaly, malformation of cortical development, basal ganglia calcifications, and mega cisterna magna [45]. These findings support the concept that ZIKV interferes with normal neuronal migration during development which then impacts the brain development. The major neuronal migration is occurring before the 25th week of gestation, making exposure to the virus in the first and second trimesters the most devastating. Infants with ZIKV exposure and no apparent congenital syndrome are also at risk for abnormal neurodevelopmental outcomes, as evidenced in a recent study of 70 infants followed to age 18 months [46]. These infants had confirmed exposure to ZIKV but no findings to support CZS, and despite the normal head circumference, had subsequent neurodevelopmental deficits develop over the first year of life [46]. As studies continue and longer-term outcomes become known, it is critically important to follow any infant with ZIKV exposure closely.

3.3 Rubella

Rubella is caused by a single stranded ribonucleic acid (RNA) virus which is highly contagious and only transmitted between humans [18, 47]. It is usually spread through respiratory droplets and in most cases will result in a mild viral disease. Symptoms may include fever, rash, malaise and adenopathy. The virus is able to infect cells of the respiratory tract and then spread via the systemic circulation to multiple organ systems, including the placenta [48]. When the infection occurs during pregnancy the virus can be transmitted to the fetus and result in death of the fetus or a range of congenital anomalies known collectively as Congenital Rubella Syndrome (CRS) [18]. The timing of when a pregnant woman contracts the virus appears to be related to the risk of congenital infection and fetal defects. Studies estimate that maternal infection occurring during the first 12 weeks of gestation has roughly a 90% chance of congenital infection with the risk of defects nearly 85% [49]. When congenital infection occurs during the first trimester, hearing defects, heart defects, neurologic damage, and ocular defects appear more commonly. CRS is a combination of these defects but most classically is described as a triad of cataracts, congenital heart disease, and sensorineural deafness [49, 50]. Other manifestations include intrauterine growth restriction (IUGR), hepatomegaly, splenomegaly, thrombocytopenia and dermal erythropoiesis (commonly known as a "blueberry muffin rash") [18].

Pregnant women in the United States are tested for rubella immunity by serologic screening. Those who have had a natural infection or have received at least one dose of the rubella vaccine tend to have lifelong immunity [18]. Those women who are found to be non-immune should receive one dose of the vaccine after childbirth, as vaccination during pregnancy has theoretical teratogenic risks due to the vaccine being live [18]. If a pregnant woman is exposed to the rubella virus, they should have serologic testing for rubella-specific IgM and IgG. If she is found to have rubella-specific IgG, then she is considered immune. However, if there is no IgG detectable at the time of exposure then convalescent serologies are obtained 3 and 6 weeks after exposure, with IgG reactivity at these time points indicating a recent infection [18]. Unfortunately, there is no treatment for rubella outside of supportive measures.

When congenital infection is suspected, diagnosis can be done by testing for rubella-specific IgM in fetal blood or detection of the virus in amniotic fluid [49].

Postnatally, an enzyme-linked immunosorbent assay (ELISA) can also be done for rubella-specific IgM. If positive, then confirmatory testing is done by reverse transcription polymerase chain reaction (RT-PCR) of nasopharyngeal swabs, urine, or oral fluid [47, 49]. In some infants the virus can be detected in nasopharyngeal secretions and urine for over a year [18, 49]. While there is no treatment for CRS, diagnosis is important in terms of follow up. Due to the risk of cataracts among other ocular abnormalities (including microphthalmia, glaucoma, chorioretinitis), hearing loss, neurologic manifestations (such as developmental delay, autism), and endocrine disorders (including diabetes, thyroid disease) children with CRS must be evaluated periodically for management of these potential complications [48–50]. The introduction of the vaccine has resulted in a significant decline in cases of rubella infection and CRS in the United States, with an average of 14 reported rubella cases a year and 4 CRS cases a year from 2001 to 2004 [51].

3.4 Cytomegalovirus

Cytomegalovirus (CMV) is a double stranded deoxyribonucleic acid (DNA) virus that is universally found and generally causes mild or subclinical symptoms in most children and adults [18, 52]. It can be transmitted via contact with infected secretions, transfusion of blood products from infected donors, organ transplants from infected individuals, or vertically [18]. When it is vertically transmitted, CMV has the potential to cause severe and permanent sequelae [18, 52, 53]. CMV is known as one of the most common congenital viral infections and is the leading, non-genetic cause of sensorineural hearing loss in children in the United States [18]. It can be transmitted to the fetus by crossing the placenta, through contact of infected cervical secretions during birth, or perinatally by ingestion of breast milk containing the virus [18]. When CMV is transmitted *in utero,* it can be due to primary maternal infection during pregnancy, reactivation of a prior infection, or reinfection with a different strain despite presence of maternal antibodies [54, 55]. Reactivation and reinfection are more common than a primary infection; however, the latter tends to cause more severe sequelae especially if infection occurs earlier in pregnancy.

Of those infants whose mother had an acute infection during pregnancy, 30–40% will have congenital CMV (cCMV) [18, 55]. Infants with cCMV are symptomatic in 10–15% of the cases, with half to two-thirds of these infants developing sensorineural hearing loss (SNHL) later in life [55]. Symptoms at birth can include thrombocytopenia, hepatomegaly, splenomegaly, microcephaly, periventricular calcifications in the brain, chorioretinitis, hepatitis, and SNHL. Long term outcomes include progressive SNHL and neurodevelopmental delay [18, 53, 55]. Of the infants who are asymptomatic at birth, around 15% will later develop SNHL [18]. Imaging of the fetal brain can be completed *in utero* via transvaginal ultrasound or with magnetic resonance imaging (MRI). cCMV can result in germinolytic cysts, lenticulostriate vasculopathy, temporal lobe and occipital cysts as well as cerebellar hypoplasia and migrational disorders including polymicrogyria [52]. Periventricular calcifications is the most frequently reported finding on brain imaging of cCMV cases, impacting 34–70% of diagnosed patients [56].

Testing during pregnancy is not routinely done, but serologic testing can be performed if a pregnant woman has been exposed or is suspected of having CMV infection. CMV-specific IgM has low specificity as it can persist for 6–9 month following primary infection and can also be detected during reactivation [54]. CMV IgG avidity index however can be used to confirm primary infection; avidity testing is a method to measure the strength of the bonding between antibodies and the virus. Low avidity would indicate recent infection while high avidity takes time to occur and would indicate a past infection. There is no current recommended treatment for acute CMV infection during pregnancy [18, 54].

There is also no current routine testing for CMV in infants. Some states have mandated targeted CMV screening for those who fail their routine newborn hearing screen, however it is important to note that targeted screening will miss those newborns who are asymptomatic at birth but still at risk for developing SNHL later in life [18]. For symptomatic infants, the diagnosis of cCMV can be made postnatally if testing is done within 3 weeks of birth as to avoid the difficulty of differentiating between intrauterine and perinatal infection [18, 54, 57]. CMV can be isolated from the urine, saliva, respiratory secretions, blood, or cerebrospinal fluid [18]. Viral cultures, rapid shell vial cultures, and PCR can be completed [54]. Treatment for those infants who are symptomatic regardless of CNS involvement includes intravenous ganciclovir or oral valganciclovir [18, 54, 58]. The latter is preferred due to ease of administration as duration of treatment is six months. If there are concerns for abnormal gastrointestinal absorption due to other factors, treatment can be started with IV ganciclovir [54]. Studies have found that those who have antiviral treatment started within the first month of life have significantly improved audiologic and neurodevelopmental outcomes at 12 and 24 months of age compared to those who do not [53]. Treatment with either valganciclovir or ganciclovir can cause significant neutropenia; absolute neutrophil counts should be monitored weekly for the first six weeks of treatment, followed by screening at eight weeks of treatment, and thereafter monthly for the duration of treatment [54]. Infants with mild symptoms or isolated SNHL are not recommended to receive antiviral treatment at this time due to lack of data in this population [54].

Long term outcomes to consider in children with cCMV include SNHL and neurodevelopmental delay. These children should have frequent audiologic assessments as SNHL can develop and/or progress after the newborn period [54]. While there are no established universal guidelines for hearing evaluation, studies indicate that screening should continue for at least the first four years of life after which late-onset SNHL is seldom seen.

3.5 Herpes simplex virus

Herpes simplex viruses are large, double-stranded DNA viruses with two types, HSV-1 and HSV-2 [18]. Traditionally, HSV-1 can cause vesicular lesions in areas above the waist while HSV-2 involves areas below the waist. It is, however, becoming increasingly more common to see genital HSV-1 lesions. Both types are able to cause herpetic disease in neonates when acquired from the mother. Transmission can occur during the birthing process via contact with genital lesions, an ascending infection, intrauterine, or postnatally from contact with lesions [18, 52]. A primary genital HSV infection in the mother near delivery has 10–30 times the risk of transmission compared to a recurrent infection. This is thought to be due to lower concentrations of transplacental HSV antibodies in the neonate [18, 59]. Unfortunately, defining an infection as primary versus recurrent may not be straightforward, as women can be asymptomatic and may be unaware that they have had a prior infection with HSV. Furthermore, viral shedding can occur in the absence of clinical symptoms [59].

If a pregnant woman does have genital lesions characteristic of HSV near delivery, then swabs of the lesions can be sent for viral culture and PCR with serologic testing to determine the type. From these results, women can be classified into four different categories: documented first primary infection, documented first episode non-primary infection, assumed first episode (primary or non-primary), or recurrent infection (see **Table 1** adapted from Kimberlin et al.).

CLASSIFICATION	CULTURE OR PCR	HSV-1 ANTIBODY	HSV-2 ANTIBODY
DOCUMENTED FIRST PRIMARY INFECTION	Positive	Negative	Negative
DOCUMENTED FIRST EPISODE NON-PRIMARY	Positive for HSV-1	Negative	Positive
	Positive for HSV-2	Positive	Negative
ASSUMED FIRST EPISODE (PRIMARY OR NON-PRIMARY)	Positive for HSV-1 or HSV-2	Unknown	Unknown
	Negative or unknown	Negative or unknown	Negative or unknown
RECURRENT INFECTION	Positive for HSV-1	Positive	Negative
	Positive for HSV-2	Negative	Positive

Table 1.
Diagnostic tests for Herpes simplex virus (HSV) Antibodies and Culture/PCR. This table describes the classification of HSV infection based on culture or PCR test results as well as HSV-1 and HSV-2 antibody test results.

Women classified as having a primary infection or first episode can be treated with oral acyclovir for 7–10 days [18]. Those with a recurrent episode can be treated with the same or higher dose for 5 days [18]. If a woman has a known history of HSV then suppressive therapy should be started at 36 weeks' gestation to decrease the risk of recurrence at delivery, although this will not entirely suppress shedding [60]. Other preventative methods include avoiding invasive fetal monitoring, such as fetal scalp electrodes, and opting for elective cesarean sections when lesions are present at the time of delivery [52, 60].

Neonatal HSV can have different manifestations. SEM disease includes disease of the skin, eyes and/or mouth; 45% of infants with HSV will have SEM. Another 30% of infants with HSV will have localized central nervous system (CNS) disease with or without skin involvement. The remaining 25% of infants with HSV will have disseminated disease which can involve multiple organs, most commonly the liver and lungs [18]. The onset of disease varies between the different manifestations, with SEM disease presenting at 5–11 days of life, CNS disease presenting between 8 and 17 days of life, and disseminated disease presenting between 10 and 12 days of life [61]. Initial symptoms may be non-specific and include feeding difficulties, lethargy, seizures, suspected sepsis, vesicular rash or severe liver dysfunction, with as many as 30% of infected neonates not having skin lesions [52, 60]. As there can be high morbidity and mortality rates in newborns with HSV, it is imperative to diagnose and initiate treatment as soon as it is suspected [18].

Guidelines have been published on the management of asymptomatic neonates born to women with active genital lesions [59]. In newborns whose mothers have a history of genital HSV prior to pregnancy and present with active lesions at delivery, there is a low risk of transmission. However, the infant should still have surface swabs of the mouth, nasopharynx, conjunctivae, and anus obtained for culture and PCR as well as serum HSV PCR sent at 24 h of life. Waiting to send samples until 24 h of life ensures that any positive results would represent active viral replication in the infant and not maternal contamination [59]. Intravenous acyclovir is not started in this situation unless the infant becomes symptomatic, or the surface swabs and/or serum are positive. This would confirm infection and require a lumbar puncture to obtain cerebrospinal fluid (CSF) for PCR testing. The result of the CSF PCR is key in determining treatment duration. If the CSF and serum HSV PCR are negative, then empiric IV acyclovir is administered for a total of 10 days to prevent progression from infection to disease. If the CSF PCR is positive, then treatment should be administered for 21 days [59]. After the treatment course has completed, a repeat lumbar puncture is necessary in cases of CNS disease to document clearance. If the repeat CSF HSV PCR is still positive, then acyclovir is continued for another 7 days. A repeat lumbar puncture is obtained to show clearance. This process is repeated until the CSF is negative. Any infant who undergoes a treatment course for HSV disease should have suppressive therapy with oral acyclovir for 6 months after the completion of parenteral treatment (see **Figures 1** and **2**) [59, 62].

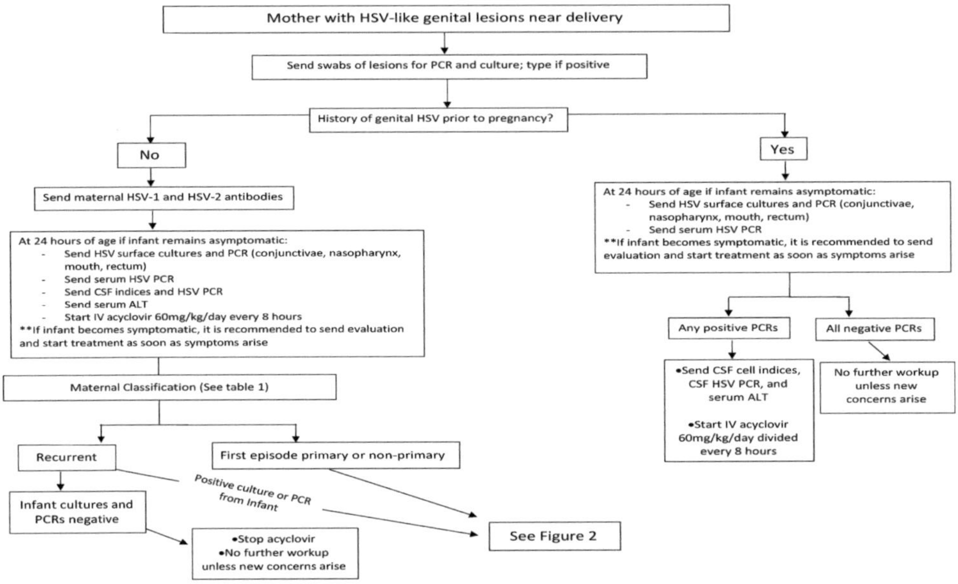

Figure 1.
Infant evaluation in suspected exposure to Herpes simplex virus (HSV). This flow diagram, adapted from Ref. [59], describes the infant evaluation(s) to complete if there was concern for maternal HSV infection around the time of delivery due to the presence of lesions.

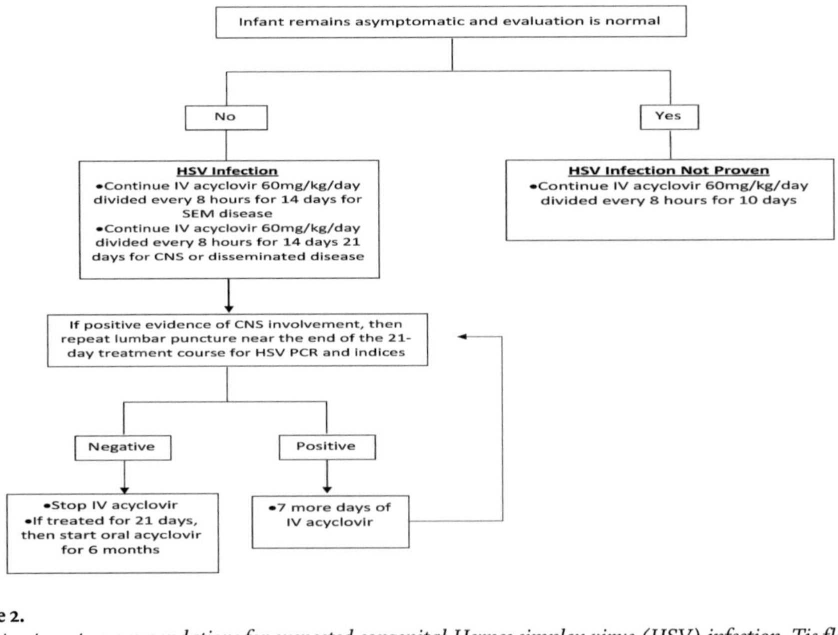

Figure 2.
Infant treatment recommendations for suspected congenital Herpes simplex virus (HSV) infection. Tis flow diagram, adapted from Ref. [59], describes treatment regimens based on infant symptoms.

In the case that an asymptomatic neonate is born to a mother with active genital lesions but does not have a history of genital HSV prior to pregnancy, then the importance lies in distinguishing whether it is a primary, non-primary or recurrent infection [59]. The mother should not only have the swabs sent for PCR testing and culture but should also have serum serological tests performed for HSV-1 and HSV-2 antibodies. The infant requires evaluation at 24 h of life with HSV surface cultures and PCR testing of the serum and CSF. The CSF samples should also be sent for cell count and chemistries, with screening serum alanine aminotransferase obtained. IV acyclovir would be started empirically after obtaining the samples at 24 h of age while awaiting results. Once the maternal testing is resulted, maternal classification can then be determined as shown in **Table 1**. If the mother is deemed to have a first episode primary or non-primary infection, then treatment of the infant would include 10 days of IV acyclovir for a normal evaluation (infant remains asymptomatic, negative CSF and serum HSV PCR, normal CSF indices, and normal serum ALT), 14 days for an abnormal evaluation (positive serum HSV PCR, symptomatic infant, or abnormal ALT) and 21 days for CNS infection (positive CSF PCR or abnormal indices) [59]. A neonate with a positive CSF HSV PCR, regardless of the maternal classification, would be managed as described above for HSV disease. It is important to note that if the infant becomes symptomatic at any point, even prior to the testing obtained at 24 h of life, then immediate evaluation and treatment should be initiated [59]. Other risk factors that may prompt testing and treatment prior the 24 h include: prolonged rupture of membranes (>4–6 h) and prematurity (≤37 weeks' gestation) in the setting of maternal genital lesions characteristic of HSV [59].

Only 10% of infants survive in untreated HSV disseminated disease with 50% of infants surviving in untreated HSV CNS disease [61]. Inadequately treated or untreated HSV SEM disease can progress to either disseminated or CNS disease; those that survive have a significant proportion that show some neurologic sequelae, namely in the form of motor, speech, and developmental delay [61]. Outcomes, especially mortality, improve the earlier that treatment is initiated, making it imperative to evaluate and begin empiric treatment whenever HSV infection is suspected [61]. Oral suppressive therapy has also been shown to improve neurodevelopmental outcomes at 12 months of age compared to those that did not receive long-term antivirals, suggesting that ongoing neurologic injury may occur in infants affected by HSV disease [62].

4. Additional viruses

A review of additional viruses that can impact infants exposed during pregnancy is provided below. These viruses have been associated with a range of adverse outcomes in infants with prenatal/perinatal exposure, however they remain uncommonly diagnosed or the impact on the fetus remains extremely varied. However, given the increased risk of potential adverse outcomes, they are briefly discussed.

4.1 Hepatitis E

The hepatitis E virus (HEV) is a single-stranded RNA virus which is known as a major cause of acute viral hepatitis especially in developing countries through ingestion of contaminated water sources [18, 63]. While it generally causes a mild illness in most adults, pregnant women tend to have more severe disease. Mortality has been observed in pregnant women, especially if infected with genotype 1 [18, 63]. HEV is estimated to be responsible for up to 3000 stillbirths a year in developing countries and can commonly cause preterm delivery in infected mothers with resultant poor neonatal outcomes [63, 64]. When HEV is transmitted vertically, hepatitis can be present from birth and persist throughout the infant's life but is not known to be associated with congenital anomalies.

4.2 Enterovirus

Enteroviruses are a group of RNA viruses that can spread between humans via respiratory routes, vertically, and fecal-oral transmission [18]. Symptoms in adults and children can be varied and may include respiratory, dermatologic, neurologic, ocular, cardiac, muscular, and gastrointestinal manifestations [18]. When enterovirus is transmitted vertically or more commonly peripartum, the neonate may remain asymptomatic without sequelae or have severe symptoms including septic shock with multiorgan dysfunction [65]. There is limited evidence to suggest that infection with enterovirus during pregnancy is associated with congenital anomalies or fetal death [65].

4.3 Congenital lymphocytic choriomeningitis virus syndrome

Lymphocytic choriomeningitis virus (LCMV) is a single-stranded RNA virus spread by rodents which can cross the placenta; rarely it can be transmitted during delivery by exposure to maternal secretions or blood and cause congenital viral infection [66–68]. Infected pregnant women can have non-specific viral symptoms and may report direct exposure to or the presence of rodents in their homes [66, 68]. Common findings in an infant affected by LCMV are macrocephaly or microcephaly and ocular abnormalities; additionally, neurological abnormalities may be present and include hydrocephalus, periventricular calcifications, seizures, neurodevelopmental sequelae including intellectual disability, or even death [67, 68]. These symptoms suggest a similarity with other congenital infections previously discussed, such as CMV or toxoplasmosis, which may contribute to an underestimation of the prevalence of LCMV when congenital infection is suspected [66, 68].

4.4 West Nile Virus

The West Nile Virus (WNV) is a flavivirus that was initially isolated in 1937 and did not reach the United States until an outbreak in 1999 [69–71]. The primary mode of transmission is through the bite of an infected *Culex* species mosquito, with individuals ranging from no symptoms to 0.7% of infected individuals developing neuro-invasive disease with encephalitis, meningitis or acute flaccid paralysis possible [69]. There is no specific treatment or vaccine at this time [70]. Case reports of infants born to mothers with WNV have shown an array of outcomes, with follow up at 2–3 years of age not consistently showing any developmental delays [69]. Findings have included chorioretinitis, white-matter loss and cystic changes, and congenital defects such as lissencephaly, polydactyly, aortic coarctation and cleft palate [69]. Additional studies on the impact of infants with exposure during gestation, and longer-term outcomes are needed to truly delineate if WNV results in congenital anomalies.

4.5 Adenovirus

Human adenoviruses (HAdV) are DNA viruses in the Adenoviridae family, with 7 subgroups and 52 serotypes [72]. While typically the cause of a "cold", the severity of illness can range from mild to severe with gastroenteritis, pneumonia and neurologic disease possible [73]. Reports have not noted any specific fetal malformations, although infants with positive polymerase chain reaction (PCR) testing had a higher incidence of neural tube defects and echogenic liver lesions with and without hydrops [74].

5. Conclusion

Many of the maternal infections that previously resulted in significant impact and poor outcomes on the developing fetus have improved as treatments and vaccines have been introduced and refined. However, other pathogens are now becoming more apparent in their impact on fetal development, such as Zika virus. Some infections are declining in incidence, with a resultant decrease in congenital infections (such as the nearly 80% decline in Rubella infections) [51]. Other infections are continuing to increase, with the true impact on society yet to be determined. Thus, it is imperative that we monitor any infections in a pregnant woman, and complete a thorough examination and evaluation of each infant born with the hopes of identifying any abnormalities quickly and improving the outcomes of each infant to the best of our ability.

Author details

Hellen Ko[1], Walter Dehority[2] and Jessie R. Maxwell[1,3]*

1 Pediatric Department, Division of Neonatology, University of New Mexico, Albuquerque, NM, United States

2 Pediatric Department, Division of Infectious Disease, University of New Mexico, Albuquerque, NM, United States

3 Neuroscience Department, University of New Mexico, Albuquerque, NM, United States

*Address all correspondence to: jrmaxwell@salud.unm.edu

References

[1] Gynecologists TACoOa. Routine Tests During Pregnancy. 2019. Available from: https://www.acog.org/Patients/FAQs/Routine-Tests-During-Pregnancy?IsMobileSet=false

[2] Prevention CfDCa. Syphilis. 2019. Available from: https://www.cdc.gov/nchhstp/pregnancy/effects/syphilis.html

[3] Kadambari S et al. Congenital viral infections in England over five decades: A population-based observational study. The Lancet Infectious Diseases. 2019;**20**:220-229

[4] Boppana SB et al. Pathogenesis of non-Zika congenital viral infections. Journal of Infectious Diseases. 2017;**216**(Suppl 10):S912-S918

[5] Tita AT, Andrews WW. Diagnosis and management of clinical chorioamnionitis. Clinics in Perinatology. 2010;**37**(2):339-354

[6] Kim CJ et al. Acute chorioamnionitis and funisitis: Definition, pathologic features, and clinical significance. American Journal of Obstetrics and Gynecology. 2015;**213**(Suppl 4):S29-S52

[7] Committee on Obstetric P. Committee opinion no. 712: Intrapartum management of intraamniotic infection. Obstetrics and Gynecology. 2017;**130**(2):e95-e101

[8] Smith MM, Daifotis HA, DeNoble AE, Dotters-Katz SK. Using the new definition of intraamniotic infection—Is there morbidity among the women left out? Journal of Maternal-Fetal and Neonatal Medicine. 2020. DOI: 10.1080/14767058.2020.1711723

[9] Gravett MG. Successful treatment of intraamniotic infection/inflammation: A paradigm shift. American Journal of Obstetrics and Gynecology. 2019;**221**(2):83-85

[10] Pappas A et al. Chorioamnionitis and early childhood outcomes among extremely low-gestational-age neonates. JAMA Pediatrics. 2014;**168**(2):137-147

[11] Bierstone D et al. Association of histologic chorioamnionitis with perinatal brain injury and early childhood neurodevelopmental outcomes among preterm neonates. JAMA Pediatrics. 2018;**172**(6):534-541

[12] Shi Z et al. Chorioamnionitis in the development of cerebral palsy: A meta-analysis and systematic review. Pediatrics. 2017;**139**(6):1-15

[13] Chau V et al. Effect of chorioamnionitis on brain development and injury in premature newborns. Annals of Neurology. 2009;**66**(2):155-164

[14] Lopez A et al. Preventing congenital toxoplasmosis. MMWR Recommendations and Reports. 2000;**49**(RR-2):59-68

[15] Khan K, Khan W. Congenital toxoplasmosis: An overview of the neurological and ocular manifestations. Parasitology International. 2018;**67**(6):715-721

[16] Sever JL et al. Toxoplasmosis: Maternal and pediatric findings in 23,000 pregnancies. Pediatrics. 1988;**82**(2):181-192

[17] Wild BM, Obringer E, Farrell E. Evaluation and treatment of fetal exposure to toxoplasmosis. NeoReviews. 2015;**16**(4):e236-e239

[18] Committee on Infectious Diseases; American Academy of Pediatrics; Kimberlin DW, Mary Anne Jackson MTB, Long SS. Red Book. 31st ed. 2018

[19] Maldonado YA, Read JS, Committee on Infectious Disease. Diagnosis,

treatment, and prevention of congenital toxoplasmosis in the United States. Pediatrics. 2017:**139**(2):e1-e51

[20] Wallon M et al. Ophthalmic outcomes of congenital toxoplasmosis followed until adolescence. Pediatrics. 2014;**133**(3):e601-e608

[21] Roizen N et al. Neurologic and developmental outcome in treated congenital toxoplasmosis. Pediatrics. 1995;**95**(1):11-20

[22] Kidd SE et al. Increased methamphetamine, injection drug, and heroin use among women and heterosexual men with primary and secondary syphilis—United States, 2013-2017. MMWR. Morbidity and Mortality Weekly Report. 2019;**68**(6):144-148

[23] Tsai S et al. Syphilis in pregnancy. Obstetrical & Gynecological Survey. 2019;**74**(9):557-564

[24] Stafford IA, Sanchez PJ, Stoll BJ. Ending congenital syphilis. JAMA. 2019;**322**(21):2073-2074. DOI: 10.1001/jama.2019.17031

[25] Schmidt R, Carson PJ, Jansen RJ. Resurgence of syphilis in the United States: An assessment of contributing factors. Infectious diseases. 2019;**12**:1178633719883282

[26] Nayak S, Acharjya B. VDRL test and its interpretation. Indian Journal of Dermatology. 2012;**57**(1):3-8

[27] Arnold SR, Ford-Jones EL. Congenital syphilis: A guide to diagnosis and management. Paediatrics & Child Health. 2000;**5**(8):463-469

[28] Peeling RW et al. Syphilis. Nature Reviews. Disease Primers. 2017;**3**:17073

[29] Hussain SA, Vaidya R. Congenital syphilis. In: StatPearls. Treasure Island, FL; 2019

[30] Trotta M et al. Epidemiology, management and outcome of varicella in pregnancy: A 20-year experience at the Tuscany Reference Centre for Infectious Diseases in Pregnancy. Infection. 2018;**46**(5):693-699

[31] Sauerbrei A, Wutzler P. The congenital varicella syndrome. Journal of Perinatology. 2000;**20**(8 Pt 1): 548-554

[32] Blumental S, Lepage P. Management of varicella in neonates and infants. BMJ Paediatrics Open. 2019;**3**(1):e000433

[33] Waring GJ. Parvovirus B19 infection: Timely diagnosis in pregnancy essential. Case Reports in Women's Health. 2018;**18**:e00057

[34] Grubman O et al. Maternal parvovirus B19 infection causing first-trimester increased nuchal translucency and fetal hydrops. Case Reports in Obstetrics and Gynecology. 2019;**2019**:3259760

[35] Crane J et al. Parvovirus B19 infection in pregnancy. Journal of Obstetrics and Gynaecology Canada. 2014;**36**(12):1107-1116

[36] Xiong YQ et al. The risk of maternal parvovirus B19 infection during pregnancy on fetal loss and fetal hydrops: A systematic review and meta-analysis. Journal of Clinical Virology. 2019;**114**:12-20

[37] Bascietto F et al. Outcome of fetuses with congenital parvovirus B19 infection: Systematic review and meta-analysis. Ultrasound in Obstetrics & Gynecology. 2018;**52**(5):569-576

[38] Vanspranghels R et al. Does an intrauterine exchange transfusion improve the fetal prognosis in parvovirus infection cases? Transfusion. 2019;**59**(1):185-190

[39] Courtier J et al. Polymicrogyria in a fetus with human parvovirus

B19 infection: A case with radiologic-pathologic correlation. Ultrasound in Obstetrics & Gynecology. 2012;**40**(5):604-606

[40] Bonvicini F et al. Gestational and fetal outcomes in B19 maternal infection: A problem of diagnosis. Journal of Clinical Microbiology. 2011;**49**(10):3514-3518

[41] MacDonald PDM, Holden EW. Zika and public health: Understanding the epidemiology and information environment. Pediatrics. 2018;**141**(Suppl 2):S137-S145

[42] Zimmerman MG, Wrammert J, Suthar MS. Cross-reactive antibodies during Zika virus infection: Protection, pathogenesis, and placental seeding. Cell Host & Microbe. 2020;**27**(1):14-24

[43] Wheeler AC. Development of infants with congenital Zika syndrome: What do we know and what can we expect? Pediatrics. 2018;**141**(Suppl 2): S154-S160

[44] Ventura CV, Ventura LO. Ophthalmologic manifestations associated with Zika virus infection. Pediatrics. 2018;**141**(Suppl 2):S161-S166

[45] Radaelli G, Lahorgue Nunes M, Bernardi Soder R, de Oliveira JM, Thays Konat Bruzzo F, Kalil Neto, F, et al. Review of neuroimaging findings in congenital Zika virus syndrome and its relation to the time of infection. The Neuroradiology Journal. 2020. DOI: 10.1177/1971400919896264

[46] Mulkey SB, Arroyave-Wessel M, Peyton C, et al. Neurodevelopmental abnormalities in children with in utero Zika virus exposure without congenital Zika syndrome. JAMA Pediatrics; 2020;**174**(3):269-276. DOI: 10.1001/jamapediatrics.2019.5204

[47] Lambert N et al. Rubella. Lancet. 2015;**385**(9984):2297-2307

[48] Mawson AR, Croft AM. Rubella Virus infection, the congenital rubella syndrome, and the link to autism. International Journal of Environmental Research and Public Health. 2019:**16**(19):1-28

[49] Bouthry E et al. Rubella and pregnancy: Diagnosis, management and outcomes. Prenatal Diagnosis. 2014;**34**(13):1246-1253

[50] Chauhan N et al. Psychiatric manifestations of congenital rubella syndrome: A case report and review of literature. Journal of Pediatric Neurosciences. 2016;**11**(2):137-139

[51] McLean HQ et al. Prevention of measles, rubella, congenital rubella syndrome, and mumps, 2013: Summary recommendations of the Advisory Committee on Immunization Practices (ACIP). MMWR Recommendations and Reports. 2013;**62**(RR-04):1-34

[52] de Vries LS. Viral infections and the neonatal brain. Seminars in Pediatric Neurology. 2019;**32**:100769

[53] James SH, Kimberlin DW. Advances in the prevention and treatment of congenital cytomegalovirus infection. Current Opinion in Pediatrics. 2016;**28**(1):81-85

[54] Tanimura K, Yamada H. Potential biomarkers for predicting congenital cytomegalovirus infection. International Journal of Molecular Sciences. 2018;**19**(12):1-13

[55] Lim Y, Lyall H. Congenital cytomegalovirus—Who, when, what-with and why to treat? The Journal of Infection. 2017;**74**(Suppl 1):S89-S94

[56] Fink KR et al. Neuroimaging of pediatric central nervous system cytomegalovirus infection. Radiographics. 2010;**30**(7):1779-1796

[57] Lopez AS et al. Intelligence and academic achievement

with asymptomatic congenital cytomegalovirus infection. Pediatrics. 2017:**140**(5):1-8

[58] Kimberlin DW et al. Valganciclovir for symptomatic congenital cytomegalovirus disease. The New England Journal of Medicine. 2015;**372**(10):933-943

[59] Kimberlin DW et al. Guidance on management of asymptomatic neonates born to women with active genital herpes lesions. Pediatrics. 2013;**131**(2):e635-e646

[60] Pinninti SG, Kimberlin DW. Neonatal herpes simplex virus infections. Seminars in Perinatology. 2018;**42**(3):168-175

[61] Harris JB, Holmes AP. Neonatal herpes simplex viral infections and acyclovir: An update. Journal of Pediatric Pharmacology and Therapeutics. 2017;**22**(2):88-93

[62] Kimberlin DW et al. Oral acyclovir suppression and neurodevelopment after neonatal herpes. The New England Journal of Medicine. 2011;**365**(14):1284-1292

[63] Krain LJ et al. Fetal and neonatal health consequences of vertically transmitted hepatitis E virus infection. The American Journal of Tropical Medicine and Hygiene. 2014;**90**(2):365-370

[64] Chaudhry SA, Verma N, Koren G. Hepatitis E infection during pregnancy. Canadian Family Physician. 2015;**61**(7):607-608

[65] Harik N, DeBiasi RL. Neonatal nonpolio enterovirus and parechovirus infections. Seminars in Perinatology. 2018;**42**(3):191-197

[66] Bonthius DJ. Lymphocytic choriomeningitis virus: An underrecognized cause of neurologic disease in the fetus, child, and adult. Seminars in Pediatric Neurology. 2012;**19**(3):89-95

[67] Enninga EAL, Theiler RN. Lymphocytic choriomeningitis virus infection demonstrates higher replicative capacity and decreased antiviral response in the first-trimester placenta. Journal of Immunology Research. 2019;**2019**:7375217

[68] Wright R et al. Congenital lymphocytic choriomeningitis virus syndrome: A disease that mimics congenital toxoplasmosis or cytomegalovirus infection. Pediatrics. 1997;**100**(1):E9

[69] Rasmussen SA et al. Studying the effects of emerging infections on the fetus: Experience with West Nile and Zika viruses. Birth Defects Research. 2017;**109**(5):363-371

[70] Krause K et al. Deletion of pregnancy zone protein and murinoglobulin-1 restricts the pathogenesis of West Nile Virus infection in mice. Frontiers in Microbiology. 2019;**10**:259

[71] Wiley CA, Chimelli L. Human Zika and West Nile virus neurological infections: What is the difference? Neuropathology. 2017;**37**(5):393-397

[72] Liao JP et al. Severe pneumonia caused by adenovirus 7 in pregnant woman: Case report and review of the literature. The Journal of Obstetrics and Gynaecology Research. 2016;**42**(9):1194-1197

[73] Ison MG, Hayden RT. Adenovirus. Microbiology Spectrum. 2016;**4**(4):1-14

[74] Baschat AA et al. Is adenovirus a fetal pathogen? American Journal of Obstetrics and Gynecology. 2003;**189**(3):758-763

Chapter 4

Approach to the Newborn with Disorders of Sex Development

Pierre Sinajon, Riyana Babul-Hirji and David Chitayat

Abstract

The birth of a baby with atypical external and/or internal genitalia is a family crisis that requires the interaction between multidisciplinary group physicians including pediatric urologists, pediatric endocrinologists, medical geneticists, genetic counsellors, gynecologists, psychologists/psychiatrists and social workers with expertise in this field. Following each of the specialists' assessment the findings, plan for investigations, the psychosocial situation and gender assignment and treatment should be reviewed among the group members prior to meeting the family. Following the group discussion the information should be presented to the parents using easy to understand language with visual aids and their questions should be answered so that they can make an informed decision regarding gender assignment, surgical options, where medically indicated, and hormone treatment. Potential for sexual relationships and fertility preservation should be discussed. The birth of a baby with abnormalities may be associated with mother/parental guilt feeling and the point that there is nothing that they did or did not do that caused the newborn's condition. Disorder of sex development (DSD) can be divided into isolated and non-isolated according to the finding on physical examination and should be further classified into abnormalities of chromosomal abnormalities, gonadal defect, internal and external genital abnormalities. Investigations should be directed by the physical examination findings and the results of the radiological, endocrine and genetic investigation including FISH analysis, microarray analysis, DNA analysis using a variety of DSD panels and, when required, whole exome/genome sequencing.

Keywords: sex development, sex differentiation, gonadal differentiation, genital abnormalities, congenital adrenal hyperplasia, androgen secretion, androgen action, androgen insensitivity, gender assignment, psychosocial status

1. Introduction

Sex development involves many autosomal and X-linked genes acting and interacting along a short duration in a precise synergy. The process of sex determination includes 4 layers:

1. Determination of the chromosome sex

2. Determination of the gonadal sex

3. Development of the internal genitalia

4. Determination of the external genitalia

DSD is a heterogeneous group of congenital abnormalities associated with atypical development of the external and internal genitalia with an estimated incidence of 1:200–1:4500 [1]. Some of the abnormalities seem to have increased in incidence along the years with hypospadias reported to reach an incidence of 1:125 newborn males and cryptorchidism reaching an incidence as high as 3% of full-term male newborns.

The nomenclature used to describe the genital abnormalities was initially descriptive, and on many occasions offensive. The advances in genetics and genomics knowledge substantially improved our understanding of the etiology of these conditions and resulted in a need to change the nomenclature. Thus, a new concept was introduced at the Chicago Consensus Conference 2005 [2] which defined the DSD as a congenital condition in which the development of chromosomal, gonadal or anatomical sex is atypical (instead of abnormal or ambiguous).

DSD can be divided into isolated or non-isolated/associated with other major abnormalities (**Figure 1**). When non-isolated they are usually associated with a variety of chromosome abnormalities, single gene or developmental field disorders which involve different body organ and systems (**Figure 2**). This review will mainly highlight the isolated/non-syndromic group of disorders of sex development.

In humans, the gonads are populated by primordial germ cells, deriving from the yolk sac wall early in week five post conception. Normally, it is the presence or absence of the of the SRY gene in the Y chromosome in the germ cells that starts the chain of events which cause the gonadal determination as testis in males. The secretion of testicular hormones in males and their attachment to their action through receptors results in normal development of the male internal and external genitalia (**Figures 3** and **4**). The absence of the SRY gene or function and the expression of genes induced by them will result in the development of ovaries. The lack of hormones produced by the testis will determine the normal formation of female internal and external genitalia (**Figures 3** and **4**).

It was previously thought that the absence of SRY gene in the gonadal ridge will result in ovarian differentiation while the existence of the SRY gene will result in

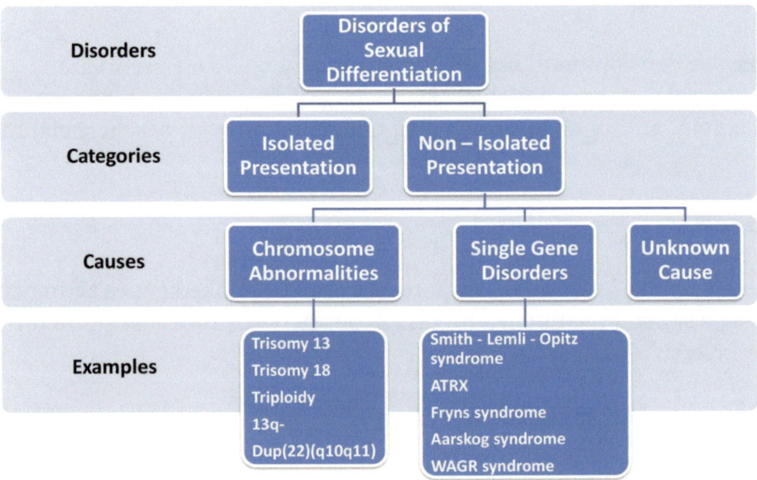

Figure 1.
Disorders of sex development—initial assessment.

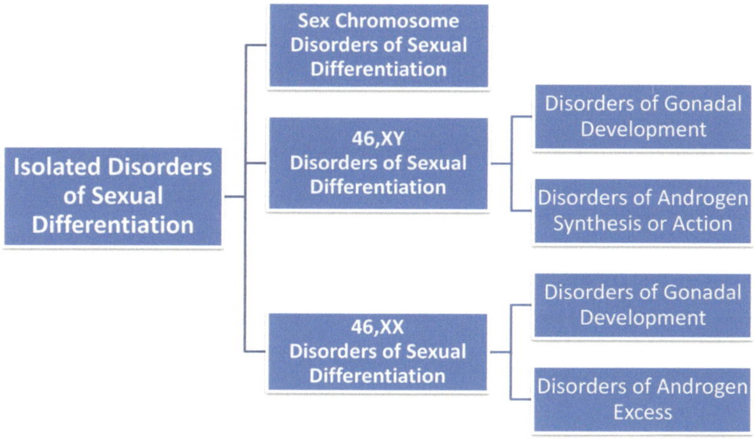

Figure 2.
Differential diagnosis in isolated disorders of sex development.

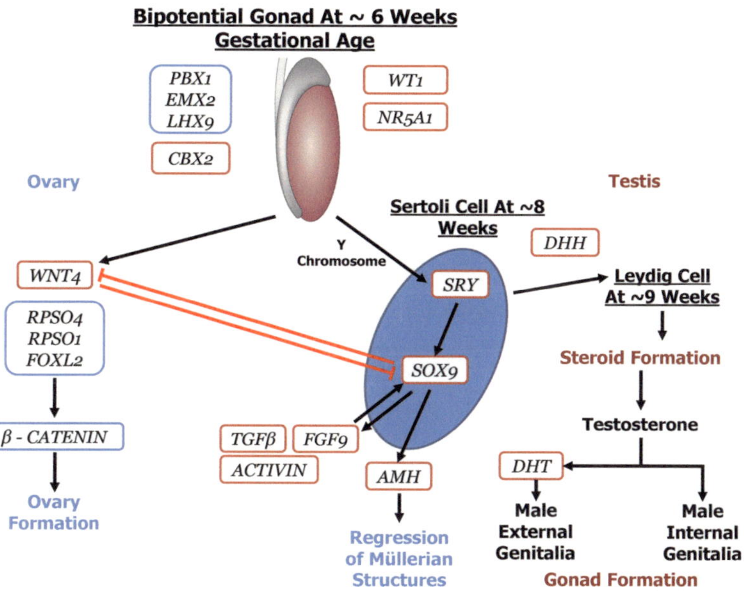

Figure 3.
Gonadal differentiation and function.

testicular development. However it is currently clear that gonadal sex is determined by antagonistic actions of ovarian and testicular cascades [5, 6] around week six post conception. In the female embryo, high levels of retinoic acid around the germ cells induces the STRA8 gene expression, resulting in germ cell meiosis and formation of oocytes. The absence of retinoic acid in the developing testis, results in formation of gonocytes which differentiate into spermatogonia and proliferate through mitosis with the meiosis happening after puberty [7].

There are three main differences between ovarian and testicular activities (**Figure 3**):

1. The testes produce testosterone early in embryogenesis which induce the formation of the male external genital while the ovaries produce hormones only from puberty

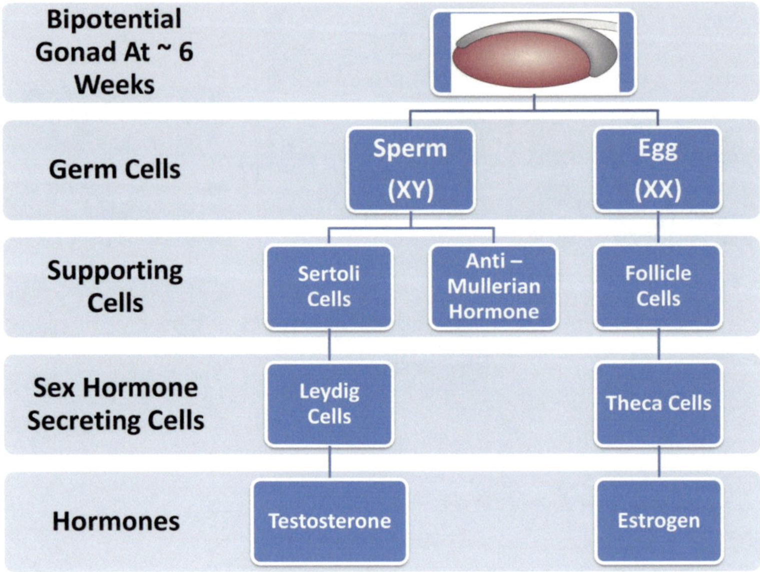

Figure 4.
Diagram of the known major genes involved in testicular and ovarian differentiation and function (adapted from [3, 4]).

2. Oogenesis starts and ends prior to 20 weeks gestation while spermatogenesis starts only at puberty

3. The ovarian hormonal production/function is linked to the process of follicle development and/or maintenance. The testes on the other hand can continue producing testosterone even in the lack of spermatogenesis.

The management of a newborn with abnormal genitalia has to be individualized taking into account the specific genital abnormalities as well as the parental and family as a whole. The decision regarding the sex of rearing should take into account the, surgical and non-surgical treatment, future pubertal development and fertility. This should be considered an urgent clinical situation and requires immediate assessment and counselling, and if possible, involvement of the DSD multidisciplinary core team including endocrinology, clinical genetics, genetic counselling, urology, obstetrics and gynecology, social work and psychology/psychiatry among others.

2. Sex chromosome disorders of sex development

Although the majority of sex chromosome abnormalities do not have genital differences as a clinical finding, some do. At conception 2% of all pregnancies are Turner syndrome and is one of the most common chromosome abnormality associated with first trimester miscarriages. The condition is highly lethal in-utero and thus the incidence at birth is 1:2500 female newborn. In almost 2/3 of cases, i.e. 60%, the karyotype is 45,X while in 15%, the condition is associated with mosaicism, i.e. 45, X/47,XXX, 45,X/46, XX/47,XXX, 45,X/46, XY, and 45,X/46, XX. Other chromosome abnormalities associated with Turner syndrome include structural abnormalities involving the X chromosome, e.g., 46,X,r(X), 46,X,Xp-, and

46,X,i(Xq) in 10% of cases, structural abnormality of the X chromosome in mosaic state in 10% of cases, with 5% in the other [8]. The condition can be detected prenatally, due to abnormalities in the lymphatic system, with the fetus presenting with heart lesions, specifically left-sided, hydrops fetalis, increased nuchal translucency or cystic hygroma. Postnatally, female babies can present with typical facial features, including epicanthic folds, droopy eyelids, down slanting palpebral fissures, micrognathia, and low set and prominent ears. The neck is often webbed and short with a nuchal hairline that is low, as well as swelling (lymphedema) of the feet and hands, and deep set nails. Left-sided cardiac anomalies are seen in 50% of the cases (hypoplastic left heart, coarctation of the aorta, bicuspid aortic valve, and aortic stenosis) with an increased risk for aortic dissection during pregnancy and the puerperal period if they decide to conceive using a donor egg. It is also associated with horseshoe kidneys as well as short stature, streak gonads and thus lack of spontaneous puberty. However, they can conceive using a donor egg.

Genetic analysis for Y chromosome material, including in a mosaic state can be picked up by chromosome microarray analysis and is important since girls with mosaicism for 46, XY [45,X/46,XY] have an increased risk for gonadoblastoma and dysgerminoma and may have absent uterus. A large British series in the literature, looking at women with Turner syndrome, showed an increased incidence not only of gonadoblastoma, but also uterine cancer, as well as possibly pediatric brain cancers. This study also demonstrated a lower incidence of breast cancer [9–11].

The incidence of Klinefelter syndrome, i.e. 47, XXY is approximately 1 in 500 males. Clinically, they can present with testes on the smaller end of the spectrum, which can impact production of testosterone and normal sperm. Otherwise, they do not have genital differences. They have normal appearing facies but can be taller in height than average and about 50% develop gynecomastia at puberty. Boys with Klinefelter syndrome can present with neurodevelopmental/neurobehavioral issues; however encouraging results, for behavioral as well as for the physical features described above, have been seen with early administration of testosterone. The optimal timing and dosage of hormonal therapy has not been established and further studies and long term follow-up are needed prior to this becoming standard treatment [12].

Mosaicism for 45,X/46, XY, also previously called mixed gonadal dysgenesis, is a rare condition with an approximate incidence reported in Denmark of less than 1/15,000 live births [8]. It has been suggested that transformation of the indifferent gonad to testes [13–15] is due to the existence of Y chromosome in the gonadal ridge. The spectrum of genital differences with this karyotype is varied, ranging from females presenting with Turner syndrome, to males with ambiguous genitalia, to phenotypically normal males. Gonadal function in most 45,X/46, XY males, even those with genital ambiguity, seems sufficient for spontaneous puberty and patients appear to benefit from GH treatment when needed [16].

3. 46, XY disorders of sex development

46, XY DSD can be divided into categories;

1. Disorders of testicular (gonadal) development

2. Disorders of androgen synthesis

Both categories lead to feminization or abnormal genitalia.

3.1 Disorders of testicular (gonadal) development

Disorders of testicular (gonadal) development are characterized by absent or small testes on palpation and/or ultrasound. Müllerian structures (uterus and fallopian tubes) can be present with the external genitalia feminized to varying degrees along with decreased levels of testosterone, dehydroepiandrosterone and androstenedione [17].

At approximately 6 weeks, the gonads can form ovaries or testis [3]. The process of testicular differentiation involves multiple genes (**Figure 4**). Mutations causing haploinsufficiency with a loss of function and duplication with gain of function are known to be associated with 46, XY gonadal DSD. The *SRY* gene encodes a transcription factor that causes a cascade effect allowing the bipotential gonad to form testis [18]. Pathogenic variants and deletions involving the *SRY* gene lead to complete gonadal dysgenesis or 46, XY pure gonadal dysgenesis. Approximately 15% of individuals with Swyer syndrome have this finding. Swyer Syndrome is also known as 46, XY complete gonadal dysgenesis or 46, XY pure gonadal dysgenesis. Patients with Swyer Syndrome present with female internal and external genitalia.

The *SRY* and *NR5A1* genes activate *SOX9* and the anti-Müllerian hormone (AMH) formation and lead to secretion by the Sertoli cells [19]. Mutations in the *SOX9* gene *[OMIM # 608160]* presents with campomelic dysplasia and sex reversal in 50% of the cases with 46, XY. A heterozygous deletion of approximately 240 kb (between 405 kb – 645 kb upstream of the *SOX9* transcription start site) was reported with 46,XY with a normal external female phenotype and severe ambiguous and asymmetric external genitalia [20]. Another case presenting as 46, XX male resulted from a heterozygous duplication upstream of the *SOX9* gene on chromosome 17 [21].

The ligand *FGF9* and the signaling molecule *WNT4* are expressed in the undifferentiated gonad. Further in development they continue to be expressed in the ovary and the testis providing opposing signals that determine gonadal differentiation. *FGF9* promotes testicular differentiation and the Wnt family member (*WNT*)/R-spondin 1 signaling and forkhead box L2 (*FOXL2*) drive female sex determination in XX gonads and promotes ovarian differentiation. Testicular formation is established when the *SRY* gene initiates a feed-forward loop. *SRY* interacts with *SOX9* and *FGF9* leading to upregulation of *FGF9* and repression of *WNT4*. The receptor for the *FGF9* in the developing testis is *FGFR2* [22]. *NR5A1* gene (also known as *SF-1*) (OMIM 184757) is another important gene for the bipotential gonad. This orphan nuclear receptor is expressed in multiple locations including hypothalamus, pituitary, gonads and adrenal glands [3]. Pathogenic variants in this gene can lead to females with premature ovarian failure. They can also present in 46,XY DSD ranging from ambiguous genitalia to female external genitalia with complete to incomplete regression of the Müllerian duct derivatives [23–25]. Various mutations including p.G35D, p.G35E, p.R92Q and p.R255L have been seen to cause adrenal insufficiency combined with gonadal dysfunction. *DAX1* is another important orphan nuclear receptor with roles in the hypothalamus, pituitary, gonads and adrenal glands [26].

The Wilms tumor factor 1 (*WT1*) pathogenic gene mutations are commonly associated with Denys–Drash and Frasier syndrome along with an increased risk for nephroblastoma. Denys–Drash classically presents with a triad of symptoms including genitourinary abnormalities, renal impairment and Wilms tumor [27]. Frasier syndrome classically presents with focal segmental glomerulonephritis (typically presenting as later onset renal impairment) and an increased risk for Wilms tumor. This risk is not as high compared to other *WT1* related conditions. Other genes of interest include *CBX2* and *DHH*. *CBX2* is involved in both the regulation of homeotic genes and of the bipotential gonad [3]. Pathogenic variants in the *CBX2* gene have presented as a normal female (internal and external genitalia). These studies suggest that *CBX2* could be responsible for repression of ovarian development [28]. In XY individuals with *MAP3K1* pathogenic variants are associated with suppressing *SOX9* and shifting the signaling pathway to promote ovarian differentiation [29]. At approximately 9 weeks of embryonic development Leydig cell differentiation occurs involving *DHH* [3]. Pathogenic variants in *DHH* can cause complete or partial gonadal dysgenesis. They can also be found in minifascicular neuropathy (OMIM 605423) [30]. Pathogenic variants in *DAX1* can cause congenital adrenal hypoplasia and hypogonadotropic hypogonadism (OMIM 300473) and complete or partial gonadal development with ambiguous external genitalia [4]. Other genes associated with XY gonadal dysgenesis include *DMRT1, DMRT2, SOX3* and *SOX8* among others [31].

3.2 Disorders of androgen synthesis

Disorders of androgen synthesis are characterized by a lack of Müllerian structures. The testes in 46, XY individuals produce Anti-Müllerian hormone and have genital abnormalities including normal female external genitalia.

Two categories can be used to describe Disorders of Androgen Synthesis:

1. Congenital adrenal hyperplasia (CAH) in patients with female chromosome sex

2. Associated with normal adrenal function.

Seven enzymes involved in the production of testosterone and dihydrotestosterone, responsible for 46, XY disorders of sex differentiation have been identified (**Figure 5**).

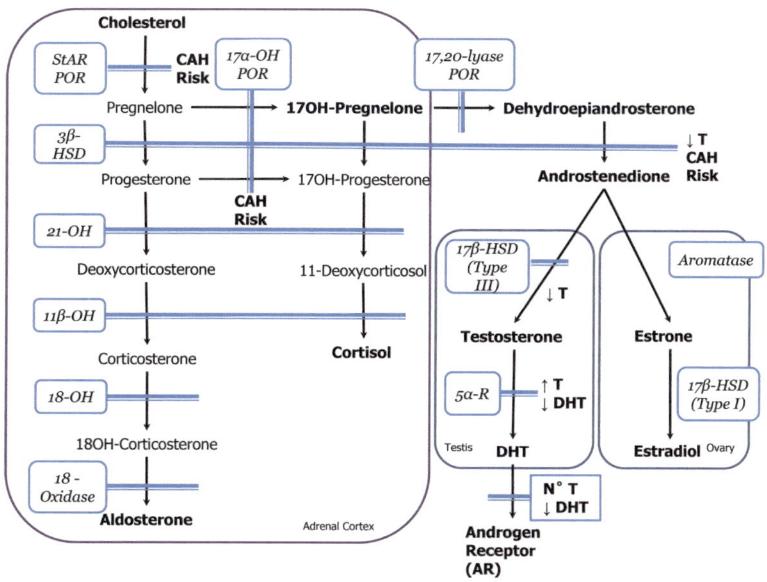

Figure 5.
Steroid hormone synthesis pathway and associated biochemical abnormalities in 46, XY DSD. POR: P450 oxidoreductase; StAR: steroid acute regulatory protein; 17α-OH: 17α-hydroxylase; 3β-HSD: 3β-hydroxysteroid dehydrogenase; 21-OH: 21-hydroxylase; 18-OH: 18-hydroxylase; 11β-OH: 11β-hydroxylase; 17β-HSD: 17β-hydroxysteroid dehydrogenase; 5α-R: 5α-reductase; CAH: congenital adrenal hyperplasia (adapted from [17]).

3.2.1 Disorders of androgen synthesis associated with adrenal dysfunction

Defects early in the pathway result in congenital adrenal hyperplasia and adrenal insufficiency. These enzymes are present in both the adrenal cortex and the gonads.

The first enzymatic step involves the steroidogenic acute regulatory (*StAR*) protein and p450 oxidoreductase (*POR*) (**Figure 5**), these enzymes cause the cleavage of cholesterol to pregnenolone. Pathogenic mutaions in the *StAR* gene lead to an autosomal recessive lipoid congenital adrenal hyperplasia (OMIM 201710). The condition is characterized by lipid accumulation, severe salt wasting and genital ambiguity. This genital involvement ranges from hypospadias to complete female external genitalia. P450 oxidoreductase defects are commonly associated with Antley-Bixler syndrome with genital anomalies and disorders in steroidogenesis (OMIM 201750). P450 oxidoreductase defects have been found with a non–syndromic presentation (OMIM 613571) with a similar picture to *StAR* mutations but lacking CAH [32].

3β-hydroxysteroid dehydrogenase (3β-HSD) is the next major enzyme involved in multiple steps including converting pregnelone to progesterone, 17α-OH pregenelone to 17α-OH progesterone and DHEA to androstenedione. Defects in 3β-HSD enzyme activity result in salt wasting and decreased testosterone production resulting in genital abnormalities in males. The findings include hypospadias, micropenis and bifid scrotum (OMIM 201810).

17α-hydroxylase is involved in the conversion pregnelone and progesterone to their 17α hydroxylated forms. Defects to this enzyme pathway also present with increased risk for adrenal insufficiency and ambiguous genitalia. Patients presenting with a combined deficiency in 17α-hydroxylase and 17,20-lyase present similarly to isolated 17α-hydroxylase deficiency. Patients with isolated 17,20-lyase deficiency have normal adrenal function and variable abnormalities of male phenotype. This is because 17,20-lyase is present in the gonads only.

3.2.2 Disorders of androgen synthesis not associated with adrenal dysfunction

17β-hydroxysteroid dehydrogenase type III (**Figure 5**) leads to the conversion of androstenedione into testosterone, this occurs within the gonads. Pathogenic variants cause an autosomal recessive disorder (OMIM 264300) with female external genitalia in 46, XY individuals, male gonadal derivatives, absent Müllerian structures, infertility and decreased testosterone [33]. These patients are sometimes difficult to distinguish from 5α-reductase deficiency and partial androgen insensitivity syndrome [34]. Further biochemical testing with ACTH stimulation or hCG stimulation may be needed [17] although DNA analysis is probably the easiest to perform.

Testosterone and dihydrotestosterone (DHT) are the end products for testis steroid hormone synthesis. Pathogenic variants occurring in the 5α-reductase gene (*SRD5A2*) lead to various biochemical changes including low levels of DHT, normal/increased levels of testosterone and high testosterone/DHT levels. Patients present with undermasculinization of the external genitalia due to low levels of DHT [34]. The presentation ranges from a female phenotype to a small phallus with severe hypospadias. These patients have normal male internal genitalia and some patients may have normal sperm production.

3.3 Disorders of androgen response

The most common cause of 46, XY DSD are the androgen insensitivity syndromes (AIS) [OMIM # 300068]. The androgen receptor (AR) is unable to activate due to the inability of testosterone or DHT to bind to the receptor [35]

(**Figure 5**). Androgens have a lack of effect on genital development. These conditions are X-linked inherited and present with a wide range in phenotypes. Complete Androgen Insensitivity Syndrome (CAIS) has an estimated prevalence of at least 1:99,000 [36] presenting with normal female genitalia and blind ending vaginal pouch. Partial Androgen Insensitivity Syndrome (PAIS) has an estimated prevalence of 1:8000. This conditions occurs when there is residual AR receptor function and hypospadias are the common finding. Mild Androgen Insensitivity Syndrome (MAIS) is the least severe. MAIS usually presents with no genital abnormalities. This condition can be suspected in the context of pubertal gynecomastia or unexplained infertility [33]. Pathogenic variants in AR have been associated with AIS. These pathogenic variants can be located outside of the coding region [37, 38]. There have been some cases where no pathogenic variants in AR have been detected. This suggests other proteins located beyond AR that influence testosterone signaling [39].

4. 46, XX disorders of sex development

46, XX DSD can occur due to:

1. Abnormal ovarian development.

2. Excess androgen levels due to abnormal synthesis or androgen exposure.

4.1 Ovarian development

There are two types of abnormal development that can cause XX sex reversal:

1. Patients who have the presence of the *SRY* gene. This can be caused by a translocation of the *SRY* gene to another chromosome, usually the X chromosome. In some rare cases a translocation to an autosome occurs.

2. Patients who are XX males and *SRY* – negative

Loss of function mutations in genes coding for ovarian formation and function are associated with ovarian dysgenesis and/or accelerated loss of primordial follicles. This can cause premature ovarian failure (POF) and/or premature menopause.

The gene responsible for the differentiation of the bipotential gonad into ovaries is the *WNT4* gene (**Figure 4**). The *WNT4* gene is a member of the WNT family of secreted molecules. This family of genes function in a paracrine manner. The WNT proteins are ligands to members of the Frizzled (FZ) family of cell surface receptors. They are also possibly ligands to the single-pass transmembrane protein LDL-receptor-related proteins 5 and 6 (*LRP5* and *LRP6*) [40]. The binding of WNT to FZ leads to reduced degradation of β-catenin. This causes β-catenin-dependent activation of T-cell factor/lymphocyte enhancer factor transcription factors that lead to the induction of WNT – responsive genes [41]. *WNT4* is produced in ovarian pre-granulosa cells. *WNT4* up-regulates the gene *DAX1* [42], which antagonizes *NR5A1*, and inhibits steroidogenic enzymes. *WNT4* – knockout XX mice have been shown to have no Müllerian ducts derivatives, have present Wolffian ducts and masculinized with the expression of the steroidogenic enzymes, namely 3β hydroxy steroid dehydrogenase and 17α hydroxylase. They are critically important in the production of testosterone. Conversely, they are normally suppressed in the developing female ovary. Mice models showed ovaries with a decreased number of oocytes. This demonstrates the important role of *WNT4* in maintaining the female germ cells and thus normal ovarian function [43]. This is contrasted by testicular function which continues in the absence of sperm.

In humans, duplication of chromosome 1p31-p35 causes a duplication of the *WNT4* gene. This duplication has been associated with male to female sex reversal exhibiting ambiguous genitalia, severe hypospadias, streak gonads and remnants of both Müllerian and Wolffian ducts [42] . Homozygotes with pathogenic variants in *WNT4* results in SERKAL (SEx Reversion, Kidneys, Adrenal and Lung dysgenesis) syndrome [44]. Pathogenic variants in this gene have also been found in women with absent Müllerian structures with clinical signs of androgen excess. Pathogenic variants have also presented with findings resembling Mayer-Rokitansky-Küster-Hauser syndrome [45].

Another important gene is the *FOXL2* gene which is responsible in the formation and function of the ovaries. Pathogenic variants in this gene result in BPES (blepharophimosis, ptosis, and epicanthus inversus) syndrome (OMIM # 110100). There are two types of the condition; in BPES I it is associated with premature ovarian failure (POF) and in BPES type II it is not associated with POF.

In mice models the continued expression of *FOXL2* is essential in maintaining ovarian function. Loss of gene expression leads to reprogramming of granulosa and theca cells into cells that are similar to Sertoli and Leydig cells, respectively [4, 46–49]. *FOXL2* also stimulates the expression of the gonadotropin releasing hormone (GnRH) receptor. This expression precedes glycoprotein hormone α-subunit, this is a common subunit to FSH, LH, and TSH in the pituitary gland [50].

RSPO1 [roof plate-specific spondin-1] is an important gene in ovarian development. It's role is to act as a regulator of female sex differentiation by activating the canonical WNT/β-catenin pathway. This pathway opposes testis formation, with *WNT4* playing a prominent role as a key ligand [51, 52]. During sex differentiation significantly higher expression of *RSPO1* was detected in the ovaries compared to testis [52], supporting the genes importance in female sex differentiation.

Other genes associated with ovarian dysgenesis and premature ovarian failure include: *LHX8*, *MCM8*, *MCM9*, *NOBOX*, and *FSHR* [46–48].

4.2 Exposure or overproduction of androgens

In 46,XX female, increased fetal androgen synthesis or prenatal exposure to androgen leads to musculinization of the female external genitalia (**Figure 6**) [26].

4.2.1 Exposure to androgens of non-fetal origin

Maternal androgen producing tumors can cause virilization of a female infant. These include adrenal tumors and ovarian tumors. There have been reports including a maternal luteoma of pregnancy which caused virilization of both mother and child [53]. There are also various drugs with androgenic activity. These include androgens, danazol, progestins and potassium sparing diuretics that are known to cause virilization [17, 53].

Placental Aromatase Deficiency is another cause of virilization in a female. This is a rare autosomal recessive condition caused by mutations in *CYP19A1* on chromosome 15q21.2 and less than 20 cases have been described in literature. Aromatase converts androstenedione to estrone in the female ovaries (**Figure 6**) providing protection for the fetus from high circulating levels of androgens. Sources of androgens include the fetal adrenal glands and androgens of maternal origin [54]. Cases present with maternal virilization in the 3rd trimester and abnormal genitalia in the affected females [54, 55]. Biochemically cases have intact cortisol and aldosterone production and are not at risk for salt wasting.

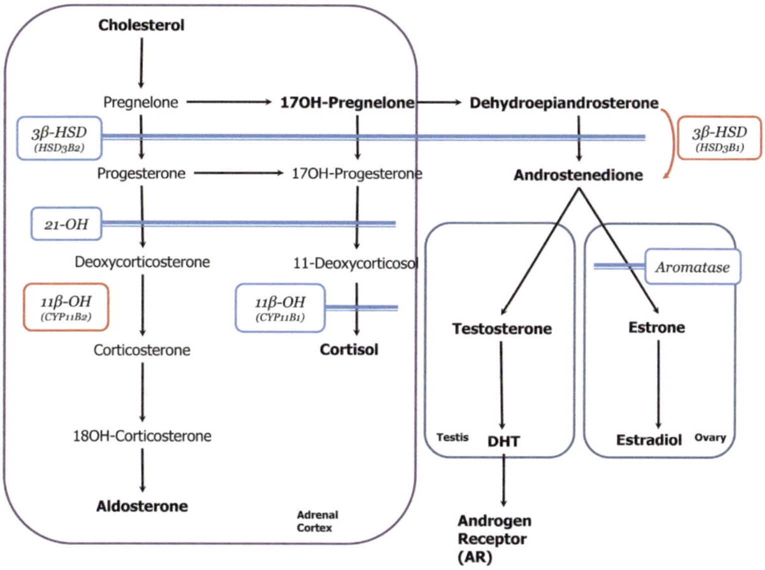

Figure 6.
Steroid hormone synthesis pathway in 46, XX DSD. 3β-HSD: 3β-hydroxysteroid dehydrogenase; 21-OH: 21-hydroxylase; 11β-OH: 11β-hydroxylase (adapted from [17]).

4.2.2 Steroid synthesis defects – overproduction of androgens

Steroid synthesis defects leading to congenital adrenal hyperplasia with androgen excess cause clitoral enlargement, partial or complete fusion of the labia majora and a short vagina [26] with virilization Prader stage above III [55].

The most common cause of 46, XX disorders of sexual differentiation (**Figure 6**) is 21-hydroxylase (21-OH) deficiency. This occurs in 90% of cases [55] and has a prevalence of 1:14,000–1:15,000 worldwide [54, 55]. This autosomal recessive condition is caused by mutations in the 21-OH gene (*CYP21A2*) on chromosome 6p21.1. The severity of the disease correlates generally to the degree of enzyme activity with classic 21(OH) deficiency having less than 5% activity. These patients present with congenital adrenal hyperplasia and in utero virilization of the external genitalia in females. Non-classic 21(OH) deficiency have >15% residual activity. These patients present with androgen excess in adolescence and early adulthood [54].

Patients with classic 21(OH) deficiency are at high risk (approximately 70%) for neonatal salt wasting [17]. These patients present with high 17α-hydroxyprogesterone, androstenedione and testosterone levels (**Figure 6**) and decreased sodium, elevated potassium and elevated renin at the end of the first week of life [17]. The high androgens levels result in virilization of the female external genitalia in female fetuses. This can be seen as early as 12 weeks gestation [54] and varies from mild clitoromegaly to complete male external genitalia with rugated and pigmented labioscrotal folds and a phallic structure [17].

Non-classic 21(OH) deficiency is more common than the classic form. The world wide incidence is 1:300, the Ashkenazi Jewish population has a higher incidence described as 1:27 [54]. They generally present in adolescence with a presentation is similar to Polycystic Ovarian Syndrome, namely premature pubarche, acne, hirsutism and irregular menses.

11β-OH is present in the adrenals and coded by the CYP11B1 gene. Defects lead to congenital adrenal hyperplasia which present with a similar picture to classical CAH. This is the second most common cause of congenital adrenal hyperplasia with an incidence of less than 1:100,000 births. The condition is inherited in an autosomal recessive manner and homozygote or compound heterozygote mutations in the *CYP11B1* gene (mapped to chromosome 8q21-q22) results in a loss of enzyme function [54]. The most common mutation in the CYP11B1 gene is the R448H mutation, which is also the founder mutation identified in Moroccan Jews who has a high prevalence of the condition [54]. Patients with 11β-OH deficiency present with virilization of the external genitalia due androgen excess; biochemically this is seen as elevated 11-deoxycortisol and 17α-hydroxypregnelone (**Figure 6**) [17]. These patients have normal aldosterone production and thus a decreased risk of salt wasting. Aldosterone synthesis is mediated by the *CYP11B2* gene, which is homologous to *CYP11B1* and located within 40 kb [54]. This results in normal levels of deoxycorticosterone and aldosterone and a wide range of blood pressure phenotypes have been reported [26, 54].

There are rare forms of defects in steroid synthesis causing CAH. These include 3β-hydroxysteroid dehydrogenase (3β-HSD) deficiency and 11β-hydroxylase (11β-OH) deficiency. There are two enzymes that mediate 3β-HSD. The type 1 enzyme (*HSD3B1*) is present in the skin, liver, peripheral tissues and placenta and the high levels of dehydroepiandrosterone are converted to androstenedione and then to testosterone and DHT [54]. The degree of virilization is moderate compared to 21-OH deficiency. Patients present with mild to moderate clitoromegaly and rarely fusion of the labioscrotal folds (**Figures 5** and **6**) [54, 55].

Type 2 (*HSD3B2*) deficiency is a rare condition occurring in 1:1,000,000 births. This condition presents with severe salt wasting similar to 21-OH deficiency and less severe virilization [54, 55]. This is an autosomal recessive condition caused by mutations in the *HSD3B2* gene, located on chromosome 1p13. As with other forms of CAH, the severity of the disease correlates with the degree of enzyme activity [54, 55]. Type 2 deficiency leads to impaired aldosterone and cortisol production presenting biochemically with elevated pregnenolone, 17(OH) pregnenolone and elevated dehydroepiandrosterone (**Figure 6**) [17, 55]. This leads to salt wasting, hyperkalemia and volume depletion similar to 21(OH) deficiency. Virilization occurs in a similar fashion to 11β (OH) deficiency.

5. Investigations

The initial investigation should include a careful physical examination to determine if the presentation is isolated or non-isolated (**Figure 1**). In view of the association between IUGR and undervirilization in chromosomal male fetuses, results of maternal serum screening, placental growth factor and detailed fetal ultrasound findings and biophysical profile as well as the birth weight should be obtained.

Examination of the external genitalia should include:

a. Assessment of the labioscrotal folds—the labioscrotal folds should be assessed for pigmentation, rugation, asymmetry and fusion. High insertion of the labioscrotal folds (above the penis) as well as "buried penis" should be differentiated from micropenis.

b. Assessment of the phallic structure, the length, breadth, urogenital openings—the phallic structure should be assessed for length, breadth, chordee and relationship with the labioscrotal folds looking for complete or incomplete penoscrotal transposition. The normal penile length at term is 3.5 cm with 2.5 cm being at −2SD. The normal clitoral length at term is 2–8.5 mm and breadth 2–6 mm [56].

c. Location, structure and volume of the gonads—gonads located in the inguinal canal and labioscrotal folds are always testes although ovotestes is also a possibility.

The perineum should also be examined for the number and position of the openings and then compared to the Prader scale [57].

Investigations should be targeted to identify:

1. The chromosome sex

2. The gonadal sex

3. The internal genitalia [using ultrasound and/or MRI to find if], is there is a uterus, is the uterus normal and are there intrabdominal gonads

4. The external genitalia

Since gonads in the inguinal canal and labioscrotal folds are almost always testis finding them on palpation usually indicates the existence of the SRY gene. However, the chromosome sex should be determined using quantitative fluorescent PCR (QF-PCR) or fluorescent in-situ hybridization (FISH) analysis. Microarray analysis should be completed, looking for submicroscopic deletion or duplication involving the *SRY* gene as well as other chromosome abnormalities [58].

Unique situations to be aware of include the most severe cases (46,XY with complete feminization and 46,XX with complete masculinization) it is difficult to diagnose an abnormal phenotype. Other scenarios include isolated grade 1 (glandular) and 2 (penile shaft) hypospadias, especially when associated with intrauterine growth restriction (IUGR) the investigative yield is low. Most cases are the result of the IUGR – Placental dysfunction – abnormal genitalia syndrome [59]; however, since this is a diagnosis by exclusion and there is no objective way of differentiating the condition from PAIS and mild 5α reductase deficiency, DNA analysis using 46,XY panel or for these conditions is recommended. Further investigations are also indicated with hypospadias associated with bilateral or unilateral undescended testes, micropenis, clitoromegaly, posterior fusion of the labia majora.

The most common cause of abnormal genitalia in female is CAH. When initial investigations show XX karyotype, 17-hydroxyprogesterone (17-OH) and renin at 48 hours of age (after the surge of adrenal hormones at birth) should be done. The abnormal sodium and potassium blood levels will present only in the second week of life.

In cases with XY karyotype, the testosterone, LH and FSH levels should be measured looking for low testosterone and dihydrotestosterone (DHT) levels. Answering the above four questions will guide further investigations which include assessment of adrenal function, testicular function and internal genitalia using ultrasound, MRI and/or genitogram or laparoscopy [17].

Genetic testing plays an important role in finding the etiology and thus providing genetic counselling regarding the recurrence risks and the prenatal/preimplantation options in future pregnancies. The use of DSD genetic panels can investigate multiple genes simultaneously and accelerate the diagnostic process. Using massive parallel sequencing looking at a variety of genes causing DSD in a cohort of 278 patients with 46, XY DSD and 48 with 46, XX DSD of an unknown etiology, Eggers et al., found a likely genetic diagnosis in 43% of patients with 46, XY DSD and 17% of patients with 46, XX DSD [60]. If no gene mutation is identified, whole exome/genome sequencing is recommended.

6. Gender assignment

When a newborn presents with ambiguous genitalia, a thorough physical examination, as well as various genetic (e.g. karyotype, microarray) and non-genetic investigations are ordered. From this, the health care team is able to have discussions with parents regarding the gender for sex of rearing. With regards to genital surgery, the current suggestion is that if genital surgery is not medically indicated, then to wait until the patient him/herself is able to make this decision. In cases with substantial ambiguity such as in PAIS, mixed gonadal dysgenesis, ovotesticular DSD and females with CAH and severe masculinization, a multidisciplinary DSD team including pediatric endocrinologists, geneticists, genetic counsellors, urologists, gynecologists, social workers and psychologists/psychiatrists with expertise in this field should be part of the care team. The findings, plan for investigations, gender assignment and treatment should be discussed among the group members prior to meeting the parents. Factors which have to be taken into consideration include type of gonads and potential function of these, present or future treatment with hormones, including for inducing puberty or for fertility if this can be predicted, as well as any medically necessary surgical treatment. When meeting with the parents, the team should present this information at a level that the parents can understand, using visual aids to facilitate the discussion. At the same time, if information is known from the literature about certain conditions that parents should be aware of, e.g. female babies with CAH and exposure to high and early levels of androgen levels prenatally can show more behaviors attributable to the male personality and sexual orientation [61–64], and the challenges in repairing severe hypospadias in patients diagnosed with PAIS [65], then this should be shared with them. In this way, the parents, together with the healthcare team's input can come to an informed decision about sex of rearing.

7. Gonadal cancer risk

Dysgenetic gonads with a Y chromosome material are associated with an increased risk for malignancy, mainly gonadoblastoma, dysgerminoma and germ cell tumors, including seminoma, non-seminoma, juvenile granulosa cell and germ cell neoplasia in situ [66, 67]. Gonadoblastoma (GB) presents with a mixture of germ cells and stromal elements as well as immature Sertoli cells and may contain calcifications with pure gonadoblastomas being not metastatic. About 80% of patients with GB are phenotypic females and 20% are phenotypic males, many of them with hypospadias and bilateral or unilateral cryptorchidism. The incidence of GB in dysgenetic gonads varies from 4.7% to as high as 25%. Germ cell neoplasia in-situ cells are found lining the seminiferous tubules in dysgenetic testes and resemble immature germ cells. Thus, they cannot be diagnosed before puberty when they are normally present. However, when found in the testes in late childhood or post puberty they are pathological and are capable of transforming into seminomas in males and dysgerminoma in females and this tumor can metastasize. Removal of the gonad prior to puberty in patients with complete androgen insensitivity is controversial in view of the data indicating low risk for malignancy until early adult years [68, 69]. In these patients, if the testes are located in the labia majora, they are amenable to ultrasound surveillance and biopsies, if needed which makes post-pubertal monitoring simpler. When the testes are undescended, laparoscopic gonadopexy to bring them near the anterior abdominal wall to allow surveillance may be an option for patients who decide to avoid gonadectomy [68, 69]. Delayed surgery can help in involving the patient in the decision making [67, 70, 71].

Persistent Mullerian Duct Syndrome is associated with the usual cancer risk associated with cryptorchidism as well as an unknown incidence of a possible tumor risk of the Mullerian duct structures.

8. Genetic counseling and DSD

In 2006, a new definition of genetic counseling was published in the Journal of Genetic Counseling, by a task force that was convened by the National Society of Genetic Counselors (NSGC). The definition is as follows: "Genetic counseling is the process of helping people understand and adapt to the medical, psychological and familial implications of genetic contributions to disease. This process integrates the following:

a. Interpretation of family and medical histories to assess the chance of disease occurrence or recurrence.

b. Education about inheritance, testing, management, prevention, resources and research.

c. Counselling to promote informed choices and adaptation to the risk or condition" [72].

Genetic counselling has an important role in the management of DSD patients and their families. In most cases, counselling involves the woman/couple, following the birth of a child with DSD and/or in discussing the implications of their future reproductive plans. Genetic counselling can also be extended to other family members and the implications to themselves and their future pregnancies. Discussions with other family members require consent from the parents of the index patient or the index patient himself/herself if of age of maturity. Other indications for genetic counselling include an adolescent or young adult with DSD who wishes to learn more about their own diagnosis and the implications for her/his reproductive plans as well a woman/couple during their pregnancy when discordance between the genotypic and phenotypic sex is identified.

Following the birth of a child with DSD, parents are often overwhelmed and confused by their child's condition. Although gender assignment and naming of the child are pressing issues for the parents, these actions should not be unduly rushed. Parents should be provided with information so as to help them in making informed decisions, together with their healthcare team. Adequate counseling and support for parents includes education regarding sexual development in utero (including brain imprinting of gender identity), genetic counseling, ethical considerations of the child's rights to make decisions regarding gender, and information regarding current guidelines and recommendations.

A three generation family history should be obtained with careful inquiry. Information about family history of infertility, women with no menstrual periods, stillbirth, recurrent miscarriages, neonatal death, congenital abnormalities, intellectual disabilities and consanguinity should be obtained. When drawing the family history, it should be kept in mind that chromosome sex and phenotypic sex may be interpreted differently by the family; for the family, it is the phenotypic sex that identifies the family members as male or female (versus chromosome sex). Furthermore, in X-linked conditions the phenotypic infertile female may be chromosomally male (such as in androgen insensitivity or ATRX).

When a diagnosis is established from genetic investigations, then the family can be provided with information regarding the clinical manifestations and natural history, mode of inheritance, medical management recommendations, if any, recurrence risk for future pregnancies, implications to their other children/other family members, as well as information regarding appropriate patient support groups. Cultural differences need to be respected and taken into consideration during the counselling process [73].

In some cases, a DSD is diagnosed incidentally during pregnancy, with no previous family history. This could be following the finding of fetal abnormal external genitalia or finding of discrepancy between the phenotypic sex as determined by fetal ultrasound and the genotypic sex as determined by chorionic villus sampling, amniocentesis or non-invasive prenatal testing (NIPT) done for other reasons. The differential diagnosis for the cases detected incidentally is broad but it should be kept in mind that in a 46,XX fetus, the most likely diagnosis is congenital adrenal hyperplasia. In a 46,XY fetus, the differential diagnosis includes a variety of conditions including androgen insensitivity syndrome, 46,XY gonadal dysgenesis and testosterone biosynthesis defects, when no other abnormalities are identified. When other abnormalities are detected, such as on ultrasound, more rare conditions such as campomelic dysplasia and Smith-Lemli-Opitz syndrome should be considered [74]. Following prenatal diagnosis of a DSD the family may choose to continue with the pregnancy with no further investigations, they may decide to terminate the pregnancy based on the information provided or they may decide to further investigate the etiology of the abnormalities identified. Prior to this however, the genetics health care team should facilitates a discussion with the family outlining each option with the pros/cons of each, including what to expect from a procedure, if applicable, as well as the impact on puberty and fertility and the potential psychosocial perspective. This information is important to allow an informed decision to be made in keeping with their value and belief system. The family should be made aware that the genetics health care team will support whatever decision they make. In situations where the family decides to terminate the pregnancy and the etiology of the condition is not known, the benefits of a fetal autopsy and molecular analysis should be reviewed with the family.

Genetic counselling should be made available to couples who are planning to have more children, following the birth of a child with DSD. The provision of genetic counselling in these cases, in addition to providing psychosocial support, includes information regarding their recurrence risk and their available reproductive options:

1. Accepting the risk of having another affected child and have no prenatal/preimplantation genetic diagnosis.

2. Deciding not to have any more biological children and choosing instead to use donor gametes, donor embryos or adoption.

3. Conceive naturally and have pre-natal diagnosis with the option of either continuing or terminating an affected pregnancy.

4. Pre-implantation genetic diagnosis (PGD).

Patients with DSD may be infertile. For these patients, conception using donor gametes and/or surrogate mother may be possible and should be discussed. However, to lower the recurrence risk, when applicable, the donor should not be a genetic relative/carrier of the condition.

9. Conclusion

Normal sex development includes determination of chromosome sex, gonadal sex, development of internal and external genitalia as well as the psychosocial sex. This is a complex process involving genetic and non-genetics components, many of them are yet unknown. Disorder of sex development is an etiologically a heterogenous group of disorders with a major lifelong impact on the patients and their families.

In view of the complexity of these group of patients they should be seen by a DSD team including but not limited to pediatric endocrinologist, pediatric urologist, medical geneticist, genetic counsellor, psychologist/psychiatrist and social worker and the findings and current knowledge should be presented to the parents so that they can make an informed decision regarding the gender, when appropriate. The child with DSD should continue be followed into puberty and adulthood to achieve optimal treatment, psychosocial well-being, sexual satisfaction, and fertility in view of the patients gender role and identity [75].

10. Resources

10.1 Web based resources

Web-based educational resources for families include:

www.aboutkidshealth.ca/En/HowTheBodyWorks/SexDevelopmentAnOverview

This website provides detailed graphically illustrated explanations of sex development and DSDs that health professionals can use when working with families.

10.2 Support groups

Androgen Insensitivity Syndrome: Differences of Sex Development Support Group.

http://aisdsd.org/

CARES foundation: Congenital Adrenal Hyperplasia Research, Education and Support.

http://www.caresfoundation.org

Hypospadias and Epispadias Association.

http://heainfo.org

Author details

Pierre Sinajon[2,3], Riyana Babul-Hirji[4] and David Chitayat[1,2,4]*

1 Department of Obstetrics and Gynecology, Mount Sinai Hospital, University of Toronto, Toronto, ON, Canada

2 The Prenatal Diagnosis and Medical Genetics Program, Department of Obstetrics and Gynecology, Mount Sinai Hospital, University of Toronto, Toronto, ON, Canada

3 Department of Pediatrics, Division of Genetics, University of Rochester Medical Center, United States

4 Division of Clinical Genetics and Metabolism, Departments of Pediatrics and Genetic Counselling, The Hospital for Sick Children, University of Toronto, Toronto, ON, Canada

*Address all correspondence to: david.chitayat@sinaihealth.ca

References

[1] Hughes IA, Nihoul-Fékété C, Thomas B, Cohen-Kettenis PT. Consequences of the ESPE/LWPES guidelines for diagnosis and treatment of disorders of sex development. Best Practice & Research. Clinical Endocrinology & Metabolism. 2007;**21**(3):351-365

[2] Hughes IA, Houk C, Ahmed SF, Lee PA. Consensus statement on management of intersex disorders. Journal of Pediatric Urology. 2006;**2**(3):148-162

[3] Rey R, Josso N, Racine C. Sexual Differentiation. In: Feingold KR, Anawalt B, Boyce A, Chrousos G, Dungan K, Grossman A, et al., editors. Endotext. South Dartmouth (MA): MDText.com, Inc. Copyright © 2000-2020, MDText.com, Inc.; 2000.

[4] Ahmed SF, Bashamboo A, Lucas-Herald A, McElreavey K. Understanding the genetic aetiology in patients with XY DSD. British Medical Bulletin. 2013;**106**:67-89

[5] Yao HH. The pathway to femaleness: current knowledge on embryonic development of the ovary. Molecular and Cellular Endocrinology. 2005;**230**(1-2):87-93

[6] Zhao F, Franco HL, Rodriguez KF, Brown PR, Tsai MJ, Tsai SY, et al. Elimination of the male reproductive tract in the female embryo is promoted by COUP-TFII in mice. Science. 2017;**357**(6352):717-720

[7] Childs AJ, Cowan G, Kinnell HL, Anderson RA, Saunders PT. Retinoic Acid signalling and the control of meiotic entry in the human fetal gonad. PLoS One. 2011;**6**(6):e20249

[8] Nielsen J, Wohlert M. Chromosome abnormalities found among 34,910 newborn children: results from a 13-year incidence study in Arhus, Denmark. Hum Genet. 1991;**87**(1):81-83

[9] Cameron-Pimblett A, La Rosa C, King TFJ, Davies MC, Conway GS. The Turner syndrome life course project: Karyotype-phenotype analyses across the lifespan. Clinical Endocrinology. 2017;**87**(5):532-538

[10] Gravholt CH, Andersen NH, Conway GS, Dekkers OM, Geffner ME, Klein KO, et al. Clinical practice guidelines for the care of girls and women with Turner syndrome: proceedings from the 2016 Cincinnati International Turner Syndrome Meeting. European Journal of Endocrinology. 2017;**177**(3):G1-g70

[11] Schoemaker MJ, Swerdlow AJ, Higgins CD, Wright AF, Jacobs PA. Cancer incidence in women with Turner syndrome in Great Britain: a national cohort study. The Lancet Oncology. 2008;**9**(3):239-246

[12] Flannigan R, Patel P, Paduch DA. Klinefelter Syndrome. The Effects of Early Androgen Therapy on Competence and Behavioral Phenotype. Sex Med Rev. 2018;**6**(4):595-606

[13] Telvi L, Lebbar A, Del Pino O, Barbet JP, Chaussain JL. 45,X/46,XY mosaicism: report of 27 cases. Pediatrics. 1999;**104**(2 Pt 1):304-308

[14] Tho SP, Jackson R, Kulharya AS, Reindollar RH, Layman LC, McDonough PG. Long-term follow-up and analysis of monozygotic twins concordant for 45,X/46,XY peripheral blood karyotype but discordant for phenotypic sex. Am J Med Genet A. 2007;143a(21):2616-22.

[15] Tosson H, Rose SR, Gartner LA. Description of children with 45,X/46,XY karyotype. Eur J Pediatr. 2012;171(3):521-9.

[16] Lindhardt Johansen M, Hagen CP, Rajpert-De Meyts E, Kjærgaard S, Petersen BL, Skakkebæk NE, et al. 45,X/46,XY mosaicism: phenotypic characteristics, growth, and reproductive function--a retrospective longitudinal study. The Journal of Clinical Endocrinology and Metabolism 2012;97(8):E1540–E1549.

[17] Wherrett DK. Approach to the Infant with a Suspected Disorder of Sex Development. Pediatric Clinics of North America. 2015;**62**(4):983-999

[18] Berta P, Hawkins JR, Sinclair AH, Taylor A, Griffiths BL, Goodfellow PN, et al. Genetic evidence equating SRY and the testis-determining factor. Nature. 1990;**348**(6300):448-450

[19] Sekido R, Lovell-Badge R. Sex determination involves synergistic action of SRY and SF1 on a specific Sox9 enhancer. Nature. 2008;**453**(7197):930-934

[20] Benko S, Gordon CT, Mallet D, Sreenivasan R, Thauvin-Robinet C, Brendehaug A, et al. Disruption of a long distance regulatory region upstream of SOX9 in isolated disorders of sex development. Journal of Medical Genetics. 2011;**48**(12):825-830

[21] Vetro A, Ciccone R, Giorda R, Patricelli MG, Della Mina E, Forlino A, et al. XX males SRY negative: a confirmed cause of infertility. Journal of Medical Genetics. 2011;**48**(10):710-712

[22] Bagheri-Fam S, Bird AD, Zhao L, Ryan JM, Yong M, Wilhelm D, et al. Testis Determination Requires a Specific FGFR2 Isoform to Repress FOXL2. Endocrinology. 2017;**158**(11):3832-3843

[23] Song Y, Fan L, Gong C. Phenotype and Molecular Characterizations of 30 Children From China With NR5A1 Mutations. Frontiers in Pharmacology. 2018;**9**:1224

[24] Tantawy S, Mazen I, Soliman H, Anwar G, Atef A, El-Gammal M, et al. Analysis of the gene coding for steroidogenic factor 1 (SF1, NR5A1) in a cohort of 50 Egyptian patients with 46,XY disorders of sex development. European Journal of Endocrinology. 2014;**170**(5):759-767

[25] Yagi H, Takagi M, Kon M, Igarashi M, Fukami M, Hasegawa Y. Fertility preservation in a family with a novel NR5A1 mutation. Endocrine Journal. 2015;**62**(3):289-295

[26] Barbaro M, Wedell A, Nordenström A. Disorders of sex development. Seminars in Fetal & Neonatal Medicine. 2011;**16**(2):119-127

[27] Firth HV, Hurst JA. Oxford Desk Reference2017.

[28] Biason-Lauber A, Konrad D, Meyer M, DeBeaufort C, Schoenle EJ. Ovaries and female phenotype in a girl with 46,XY karyotype and mutations in the CBX2 gene. American Journal of Human Genetics 2009;84(5):658-663.

[29] Loke J, Pearlman A, Radi O, Zuffardi O, Giussani U, Pallotta R, et al. Mutations in MAP3K1 tilt the balance from SOX9/FGF9 to WNT/beta-catenin signaling. Human molecular genetics. 2014;**23**(4):1073-1083

[30] Sugie K, Futamura N, Suzumura A, Tate G, Umehara F. Hereditary motor and sensory neuropathy with minifascicle formation in a patient with 46XY pure gonadal dysgenesis: a new clinical entity. Annals of Neurology. 2002;**51**(3):385-388

[31] Yatsenko SA, Witchel SF. Genetic approach to ambiguous genitalia and disorders of sex development: What clinicians need to know. Seminars in Perinatology. 2017;**41**(4):232-243

[32] Hershkovitz E, Parvari R, Wudy SA, Hartmann MF, Gomes LG,

Loewental N, et al. Homozygous mutation G539R in the gene for P450 oxidoreductase in a family previously diagnosed as having 17,20-lyase deficiency. The Journal of Clinical Endocrinology and Metabolism. 2008;**93**(9):3584-3588

[33] Yang Z, Ye L, Wang W, Zhao Y, Wang W, Jia H, et al. 17beta-Hydroxysteroid dehydrogenase 3 deficiency: Three case reports and a systematic review. The Journal of Steroid Biochemistry and Molecular Biology. 2017;**174**:141-145

[34] Massanyi EZ, Dicarlo HN, Migeon CJ, Gearhart JP. Review and management of 46,XY disorders of sex development. Journal of Pediatric Urology. 2013;**9**(3):368-379

[35] Mongan NP, Tadokoro-Cuccaro R, Bunch T, Hughes IA. Androgen insensitivity syndrome. Best Practice & Research. Clinical Endocrinology & Metabolism. 2015;**29**(4):569-580

[36] Boehmer AL, Brinkmann O, Brüggenwirth H, van Assendelft C, Otten BJ, Verleun-Mooijman MC, et al. Genotype versus phenotype in families with androgen insensitivity syndrome. The Journal of Clinical Endocrinology and Metabolism. 2001;**86**(9):4151-4160

[37] Hornig NC, de Beaufort C, Denzer F, Cools M, Wabitsch M, Ukat M, et al. A Recurrent Germline Mutation in the 5'UTR of the Androgen Receptor Causes Complete Androgen Insensitivity by Activating Aberrant uORF Translation. PLoS One. 2016;**11**(4):e0154158

[38] Känsäkoski J, Jääskeläinen J, Jääskeläinen T, Tommiska J, Saarinen L, Lehtonen R, et al. Complete androgen insensitivity syndrome caused by a deep intronic pseudoexon-activating mutation in the androgen receptor gene. Scientific Reports. 2016;**6**:32819

[39] Hornig NC, Ukat M, Schweikert HU, Hiort O, Werner R, Drop SL, et al. Identification of an AR Mutation-Negative Class of Androgen Insensitivity by Determining Endogenous AR Activity. The Journal of Clinical Endocrinology and Metabolism. 2016;**101**(11):4468-4477

[40] He X, Semenov M, Tamai K, Zeng X. LDL receptor-related proteins 5 and 6 in Wnt/beta-catenin signaling: arrows point the way. Development. 2004;**131**(8):1663-1677

[41] Nusse R. Wnts and Hedgehogs: lipid-modified proteins and similarities in signaling mechanisms at the cell surface. Development. 2003;**130**(22):5297-5305

[42] Jordan BK, Mohammed M, Ching ST, Délot E, Chen XN, Dewing P, et al. Up-regulation of WNT-4 signaling and dosage-sensitive sex reversal in humans. American Journal of Human Genetics. 2001;**68**(5):1102-1109

[43] McLaren A. Meiosis and differentiation of mouse germ cells. Symposia of the Society for Experimental Biology. 1984;**38**:7-23

[44] Mandel H, Shemer R, Borochowitz ZU, Okopnik M, Knopf C, Indelman M, et al. SERKAL syndrome: an autosomal-recessive disorder caused by a loss-of-function mutation in WNT4. American Journal of Human Genetics. 2008;**82**(1):39-47

[45] Biason-Lauber A, Konrad D. WNT4 and sex development. Sexual Development. 2008;**2**(4-5):210-218

[46] Desai S, Wood-Trageser M, Matic J, Chipkin J, Jiang H, Bachelot A, et al. MCM8 and MCM9 Nucleotide Variants in Women With Primary Ovarian Insufficiency. The Journal of Clinical Endocrinology and Metabolism. 2017;**102**(2):576-582

[47] Katari S, Wood-Trageser MA, Jiang H, Kalynchuk E, Muzumdar R, Yatsenko SA, et al. Novel Inactivating Mutation of the FSH Receptor in Two Siblings of Indian Origin With Premature Ovarian Failure. The Journal of Clinical Endocrinology and Metabolism. 2015;**100**(6):2154-2157

[48] Ren Y, Suzuki H, Jagarlamudi K, Golnoski K, McGuire M, Lopes R, et al. Lhx8 regulates primordial follicle activation and postnatal folliculogenesis. BMC Biology. 2015;**13**:39

[49] Uhlenhaut NH, Jakob S, Anlag K, Eisenberger T, Sekido R, Kress J, et al. Somatic sex reprogramming of adult ovaries to testes by FOXL2 ablation. Cell. 2009;**139**(6):1130-1142

[50] Ellsworth BS, Egashira N, Haller JL, Butts DL, Cocquet J, Clay CM, et al. FOXL2 in the pituitary: molecular, genetic, and developmental analysis. Molecular Endocrinology. 2006;**20**(11):2796-2805

[51] Pannetier M, Chassot AA, Chaboissier MC, Pailhoux E. Involvement of FOXL2 and RSPO1 in Ovarian Determination, Development, and Maintenance in Mammals. Sexual Development. 2016;**10**(4):167-184

[52] Tomaselli S, Megiorni F, Lin L, Mazzilli MC, Gerrelli D, Majore S, et al. Human RSPO1/R-spondin1 is expressed during early ovary development and augments beta-catenin signaling. PLoS One. 2011;**6**(1):e16366

[53] Spitzer RF, Wherrett D, Chitayat D, Colgan T, Dodge JE, Salle JL, et al. Maternal luteoma of pregnancy presenting with virilization of the female infant. Journal of Obstetrics and Gynaecology Canada. 2007;**29**(10):835-840

[54] Auchus RJ, Chang AY. 46,XX DSD: the masculinised female. Best Practice & Research. Clinical Endocrinology & Metabolism. 2010;**24**(2):219-242

[55] Paris F, Gaspari L, Philibert P, Maïmoun L, Kalfa N, Sultan C. Disorders of sex development: neonatal diagnosis and management. Endocrine Development. 2012;**22**:56-71

[56] Oberfield SE, Mondok A, Shahrivar F, Klein JF, Levine LS. Clitoral size in full-term infants. American Journal of Perinatology. 1989;**6**(4):453-454

[57] Allen L. Disorders of sexual development. Obstetrics and Gynecology Clinics of North America. 2009;**36**(1):25-45

[58] Tannour-Louet M, Han S, Corbett ST, Louet JF, Yatsenko S, Meyers L, et al. Identification of de novo copy number variants associated with human disorders of sexual development. PLoS One. 2010;**5**(10):e15392

[59] Yinon Y, Kingdom JC, Proctor LK, Kelly EN, Salle JL, Wherrett D, et al. Hypospadias in males with intrauterine growth restriction due to placental insufficiency: the placental role in the embryogenesis of male external genitalia. Am J Med Genet A. 2010;152a(1):75-83.

[60] Eggers S, Sadedin S, van den Bergen JA, Robevska G, Ohnesorg T, Hewitt J, et al. Disorders of sex development: insights from targeted gene sequencing of a large international patient cohort. Genome Biology. 2016;**17**(1):243

[61] Mathews GA, Fane BA, Conway GS, Brook CG, Hines M. Personality and congenital adrenal hyperplasia: possible effects of prenatal androgen exposure. Hormones and Behavior. 2009;**55**(2):285-291

[62] Nordenström A, Servin A, Bohlin G, Larsson A, Wedell A.

Sex-typed toy play behavior correlates with the degree of prenatal androgen exposure assessed by CYP21 genotype in girls with congenital adrenal hyperplasia. The Journal of Clinical Endocrinology and Metabolism. 2002;**87**(11):5119-5124

[63] Frisén L, Nordenström A, Falhammar H, Filipsson H, Holmdahl G, Janson PO, et al. Gender role behavior, sexuality, and psychosocial adaptation in women with congenital adrenal hyperplasia due to CYP21A2 deficiency. The Journal of Clinical Endocrinology and Metabolism. 2009;**94**(9):3432-3439

[64] Meyer-Bahlburg HF, Dolezal C, Baker SW, New MI. Sexual orientation in women with classical or non-classical congenital adrenal hyperplasia as a function of degree of prenatal androgen excess. Archives of Sexual Behavior. 2008;**37**(1):85-99

[65] Lucas-Herald A, Bertelloni S, Juul A, Bryce J, Jiang J, Rodie M, et al. The Long-Term Outcome of Boys With Partial Androgen Insensitivity Syndrome and a Mutation in the Androgen Receptor Gene. The Journal of Clinical Endocrinology and Metabolism. 2016;**101**(11):3959-3967

[66] Cools M, Looijenga L. Update on the Pathophysiology and Risk Factors for the Development of Malignant Testicular Germ Cell Tumors in Complete Androgen Insensitivity Syndrome. Sexual Development. 2017;**11**(4):175-181

[67] Wünsch L, Holterhus PM, Wessel L, Hiort O. Patients with disorders of sex development (DSD) at risk of gonadal tumour development: management based on laparoscopic biopsy and molecular diagnosis. BJU Int. 2012;110(11 Pt C):E958-65.

[68] Chaudhry S, Tadokoro-Cuccaro R, Hannema SE, Acerini CL, Hughes IA. Frequency of gonadal tumours in complete androgen insensitivity syndrome (CAIS): A retrospective case-series analysis. J Pediatr Urol. 2017;13(5):498.e1-.e6.

[69] Döhnert U, Wünsch L, Hiort O. Gonadectomy in Complete Androgen Insensitivity Syndrome: Why and When? Sexual Development. 2017;**11**(4):171-174

[70] Mouriquand PD, Gorduza DB, Gay CL, Meyer-Bahlburg HF, Baker L, Baskin LS, et al. Surgery in disorders of sex development (DSD) with a gender issue: If (why), when, and how? Journal of Pediatric Urology. 2016;**12**(3):139-149

[71] Patel V, Casey RK, Gomez-Lobo V. Timing of Gonadectomy in Patients with Complete Androgen Insensitivity Syndrome-Current Recommendations and Future Directions. Journal of Pediatric and Adolescent Gynecology. 2016;**29**(4):320-325

[72] Resta R, Biesecker BB, Bennett RL, Blum S, Hahn SE, Strecker MN, et al. A new definition of Genetic Counseling: National Society of Genetic Counselors' Task Force report. Journal of Genetic Counseling. 2006;**15**(2):77-83

[73] Chitayat D, Babul-Hirji R. Genetic counselling in prenatally diagnosed non-chromosomal fetal abnormalities. Current Opinion in Obstetrics & Gynecology. 2000;**12**(2):77-80

[74] Chitayat D, Glanc P. Diagnostic approach in prenatally detected genital abnormalities. Ultrasound in Obstetrics & Gynecology. 2010;**35**(6):637-646

[75] Raveenthiran V. Neonatal Sex Assignment in Disorders of Sex Development: A Philosophical Introspection. J Neonatal Surg. 2017;**6**(3):58

Chapter 5

Congenital Anomalies of the Gastrointestinal Tract

Richa Verma

Abstract

The gastrointestinal system anomalies in the newborn infants are not uncommon and are due to either embryogenesis defects or intrauterine vascular accidents resulting in a compromise of fetal blood circulation to specific system organs. The symptoms generally present within first 1–2 days of life and are mostly referable to intestinal obstruction, manifesting as vomiting, feeding difficulty and distension of abdomen. Other defects may have distinct signs and symptoms and variable time of onset. Some defects may be diagnosed antenatally during prenatal maternal care. The investigations include radiography, magnetic resonance imaging and ultrasound and in a majority of cases clinical presentation and plain radiography may provide adequate diagnostic information. The outcomes of surgical repair are variable and depend upon the original pathological defect. Here, the common anomalies are described with their clinical presentation, surgical repair techniques and outcomes.

Keywords: congenital gastrointestinal anomalies, neonates, embryogenesis, surgical repair, intestine, gall bladder, esophagus, colon, liver

1. Introduction

Congenital malformations involving the gastrointestinal tract (GIT) can be broadly divided into upper and lower gut abnormalities (**Table 1**). Upper pathology involves the foregut tubes, which are proximal to the ligament of Treitz: the esophagus, stomach, duodenum, pancreas and hepatobiliary tract. Lower GIT anomalies include the mid and hindgut structures: the jejunum and ileum, which constitute the small bowel, the colon and anorectal malformations. Congenital anomalies can further be classified based on whether the defect is structural or functional. Structural anomalies result from either defective embryogenesis or intrauterine complications, such as ischemia. Functional defects have normal anatomy but disrupted flow of GIT contents. In most cases, structural defects adversely impact functional capability. This chapter reviews the clinical presentation, diagnostic work up and surgical management of upper and lower GIT congenital anomalies.

	Anatomic relation	Embryonic source	Blood supply	Viscera
Upper gastrointestinal tract	Proximal to ligament of Treitz	Foregut	Celiac axis	Esophagus Stomach Duodenum Biliary ducts Liver Pancreas
Lower gastrointestinal tract	Distal to ligament of Treitz	Midgut	SMA	Jejunum Ileum Cecum Ascending colon Proximal 2/3 transverse colon
		Hindgut	IMA	Distal 1/3 transverse colon Descending colon Sigmoid colon Rectum Anal canal

SMA: superior mesenteric artery; IMA: inferior mesenteric artery.

Table 1.
Embryologic derivates of the gastrointestinal tract.

2. Foregut disorders

2.1 Esophageal atresia (EA) +/− tracheal fistula (TEF)

2.1.1 Embryology

During the fourth week of gestation, the embryonic ventral foregut differentiates into the esophagus and trachea. Muscular and neurovascular development of the esophagus is complete by the end of ninth week of gestation. It is likely that esophageal malformations result from errors during this developmental time period.

2.1.2 Clinical presentation

EA/TEF is categorized into five types and clinical presentation varies depending on the type of pathology (**Figure 1**). Type A is the most common (90% cases) and consists of proximal EA with a distal TEF. Type B consists solely of proximal EA (no fistula) whereas type C only has a TEF (no atresia). Type D has both a proximal and distal TEF in the setting of atresia. Type E consists of proximal EA with TEF and a distal esophageal pouch. Types D and E are exceedingly rare.

The infant will exhibit drooling and attempts at feeding will result in coughing, choking and regurgitation. Since types B and E have a proximal obstruction without distal fistulization, the infant will have a scaphoid abdomen and gas will not be seen in the bowel distally on radiograph. Type C may present with recurrent aspiration pneumonia and may not be diagnosed until later in life.

2.1.3 Diagnosis

Prenatal ultrasound will demonstrate polyhydramnios and the blind end of the esophageal pouch may be visualized. After birth, unsuccessful attempt at passage

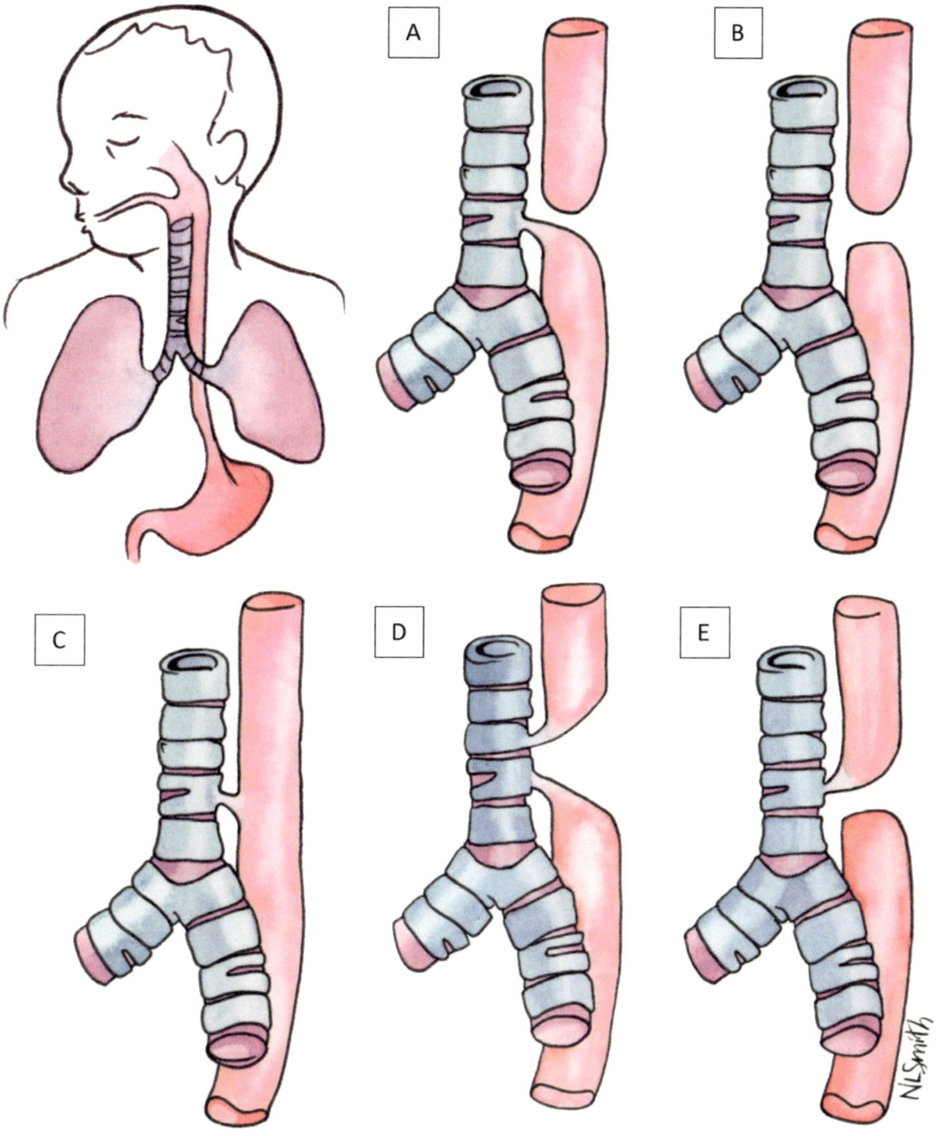

Figure 1.
Types of tracheoesophageal fistulae depicted as figures A-E.

of an oro- or nasogastric tube is diagnostic. The tip of the tube will be seen in the esophageal pouch on radiography.

Because of the VACTERL phenomenon (vertebral, anal, cardiac, tracheoesophageal, renal and limb deformities), renal and cardiac ultrasounds as well as plains films of the spine and limbs must be obtained to determine the presence of any other anomalies. An echocardiogram is particularly essential to ensure that the aortic arch is in its normal left-sided anatomic location because this impacts operative planning. Ventricular septal defect is the most common anomaly associated with EA/TEF.

2.1.4 Surgical management

Ideally, EA/TEF is corrected in a single procedure. Staged procedure, beginning with decompressive gastrostomy and fistula takedown, followed by esophageal reconstruction at a later date, is reserved for those too unstable to tolerate general anesthesia due to respiratory or cardiac defects. Infants with long gap atresia also undergo delayed repair to allow elongation of the proximal and distal esophageal ends.

In current practice, the minimally invasive approach using video assisted thoracoscopy is preferred to open thoracotomy. If the open approach is employed, a right posterolateral thoracotomy incision is made at the fourth intercostal space, sparing the serratus anterior and latissimus dorsi muscles. Extrapleural dissection is carried until the azygous vein is encountered, which is then divided. In the case of type A, the lower esophageal pouch and its associated fistula are identified. The fistula is resected. The proximal esophageal pouch is then mobilized to establish tension free continuity between the two ends. If a proximal fistula is present, this is ligated prior to mobilization. The esophagus is reconstructed via a single layer end-to-end anastomosis. A chest tube is placed and remains until post-operative esophogram confirms patency of the anastomosis. Anastomotic leaks tend to heal without intervention and are managed by continuation of chest tube and antibiotics.

2.1.5 Outcomes

Thoracoscopic approach has led to improved outcomes and most infants grow to lead fairly normal lives, given the lack of concurrent anomalies such as cardiac defects. Most commonly, gastroesophageal reflux (GER) and esophageal strictures are lifelong issues endured by the patient. GER may be asymptomatic or lead to persistent cough, respiratory problems or esophageal stricturing. Primary management is medical with anti-reflux medications and prokinetics. Surgical correction of GER with fundoplication is last resort. Esophageal strictures may form many years after repair and are best managed by endoscopic dilation. Recurrent or refractory esophageal strictures require surgical resection and re-anastomosis.

2.2 Duodenal atresia

2.2.1 Embyrology

It results due to the failure of duodenal recanalization and most commonly occurs in the second portion of the duodenum distal to ampulla of Vater but any segment can be affected.

2.2.2 Clinical presentation

Emesis and feeding intolerance occurs in the first 24–48 h of life. The type of emesis—bilious versus non—depends on the location of atresia relative to the major duodenal papilla. If obstruction is distal to it, infant will exhibit bilious emesis. Obstruction proximal to the ampulla causes non-bilious emesis. Abdomen will not be distended due to proximal nature of obstruction. A palpable mass in the epigastrium may be appreciated on physical exam.

2.2.3 Diagnosis

The "double bubble" on abdominal x-ray indicates air in stomach and duodenum but not in distal small bowel and colon. An UGI series must be obtained to rule out malrotation, which can also present with bilious emesis early in life and is a surgical emergency. UGI may reveal a duodenal web, which is an intraluminal diverticulum that appears as an elongated, conical silhouette resembling a "windsock". Echocardiogram and renal ultrasound are performed to rule out any other defects as there is an association with trisomy 21 and its related complications.

2.2.4 Surgical management

"The diamond D", Diamond Duodenoduodenostomy—A transverse incision is made in the proximal widened duodenum and a longitudinal incision in the distal tapered portion of the duodenum (**Figure 2**). The anastomosis is created in a diamond shape to facilitate mucosal abutment between the two incongruent duodenal diameters. During repair, evaluation for duodenal web must be performed because they are not always identified on pre-operative UGI and can cause persistent obstruction if not corrected. If present, a longitudinal duodenotomy is performed over the area of the web and it is excised. Careful attention must be paid to its location relative to the major duodenal papilla so as to not disrupt the integrity of the ampulla of Vater. The duodenotomy is closed in a transverse fashion to avoid narrowing of the lumen.

2.2.5 Outcomes

There tend to be few, if any, long term complications following correction of duodenal atresia. Persistent obstruction may indicate missed duodenal web and requires re-operation. Delayed gastric emptying may occur in the early postoperative period and does not warrant any intervention; most cases resolve with time and enteral feedings can be advanced in small volumes as tolerated.

2.3 Pyloric stenosis

2.3.1 Embryology

The exact etiology is unknown. Exposure to erythromycin has been implicated as a risk factor [1].

2.3.2 Clinical presentation

It is characterized with feeding intolerance and non-bilious emesis that becomes projectile over time; usually presenting around 2–4 weeks of life, however, may not present up until 6–12 weeks. Emesis is non-bilious because the site of obstruction, the pylorus, is proximal to the ampulla of Vater. It tends to occur in first born Caucasian males.

2.3.3 Diagnosis

On physical exam, may be able to palpate an "olive like" firm, mobile mass in the right upper quadrant or epigastrium, however this is often difficult to appreciate on a restless infant. Abdomen is otherwise soft and non-distended. Ultrasound is diagnostic and demonstrates a pyloric channel length $\geq$ 16 mm, wall $\geq$4 mm in thickness.

Repeated vomiting of gastric acid (HCl) leads to hypochloremia, alkalosis and dehydration. Hypovolemia stimulates aldosterone secretion with resultant sodium resorption and potassium secretion. Thus, the infant's laboratory panel will reveal hypochloremic, hypokalemic metabolic alkalosis. Hydrogen is shifted extracellularly in exchange for potassium to correct the acid–base imbalance, exacerbating hypokalemia. Eventually, worsening hypokalemia stimulates the renal hydrogen-potassium pump to resorb potassium and secrete hydrogen, resulting in acidic urine. This is termed "paradoxical aciduria" because bicarbonate secretion should take precedence in an alkalotic state, but the nephrons prioritize correction of potassium at the expense of hydrogen loss instead.

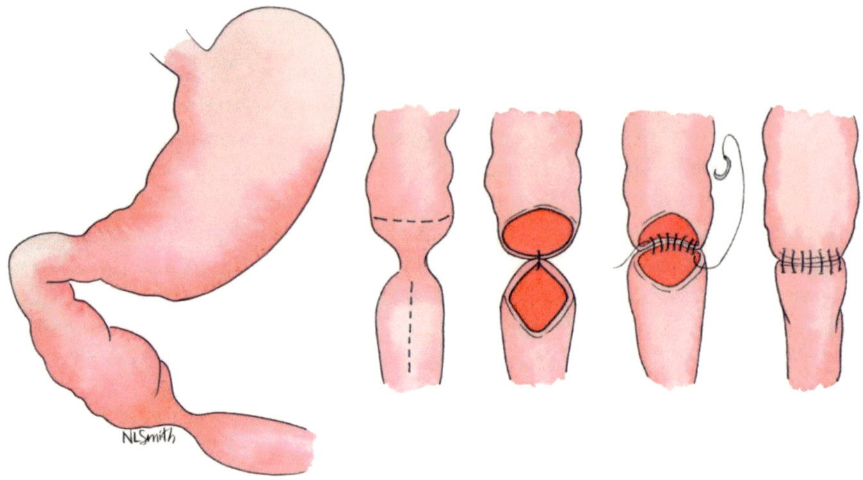

Figure 2.
Diamond duodenoduodenostomy for duodenal atresia repair.

2.3.4 Surgical management

Pyloric stenosis is not a surgical emergency and operative intervention is deferred until electrolytes have normalized, ideally, chloride >95, bicarbonate <30. As the primary metabolic derangements are caused by volume and gastric juice loss, resuscitation should be initiated with 10-20 cc/kg normal saline boluses. Once volume status has been adequately restored and urine output robust, potassium containing fluids (D5 1/2NS + 10 K/L) are administered at maintenance rate.

The Ramstedt pyloromytomy was historically carried out through a right subcostal transverse incision however the laparoscopic approach is becoming preferred in current practice. A longitudinal incision along the anterior surface of the pylorus is carried down through the serosa and hypertrophied muscle until the submucosa protrudes, much like slicing the tough outer skin of a grape until the smooth inner flesh is encountered. The length of the myotomy extends from the antrum of the stomach proximally to the pyloric vein of Mayo distally, which designates the junction of the pylorus and proximal duodenum. Oral feeding may be initiated 6–8 h post-operatively and advanced as tolerated.

2.3.5 Outcomes

Long term results from pyloromyotomy are excellent and few infants, if any, have residual complications. Incomplete myotomy can present with persistent feeding intolerance in the peri-operative period and requires re-operation.

2.4 Biliary atresia

2.4.1 Embryology

The pathophysiology is unknown. Between 4 and 10 weeks of gestation, the extrahepatic biliary tract develops from the hepatic diverticulum. This occurs normally. In the post-natal period, there appears to be an inflammatory process that causes fibrosis of the extrahepatic biliary ducts [2].

2.4.2 Clinical presentation

Worsening jaundice unamenable to phototherapy during the first 2 weeks of life, subsequently demonstrating unrelenting direct hyperbilirubinemia are characteristic. Laboratory values are consistent with biliary obstruction and demonstrate direct hyperbilirubinemia and elevated alkaline phosphatase. Signs of cholestasis, dark urine and light or gray colored stools are present.

2.4.3 Diagnosis

Hepatobiliary technetium-99 iminodiacetic acid scan (99-Tc IDA) has highest sensitivity and specificity [2]. Normally, the radiotracer is taken up by hepatocytes and readily excreted into the intestines via the biliary ducts. In biliary atresia, technetium will be taken up by the liver normally, but obstruction of the extrahepatic ducts prevents outflow of radiotracer into the duodenum. Abdominal ultrasound may reveal a small or obliterated gallbladder. Magnetic resonance cholangiopancreatography (MRCP) is also be helpful in ruling out intrahepatic atresia or choledocal cysts.

2.4.4 Surgical management

Expeditious operative intervention is imperative as liver damage can be attenuated, even reversed, and chance of survival improved with early biliary decompression. Beyond 3–4 months, irreversible liver damage may preclude successful outcome. The Kasai portoenterostomy is the procedure of choice. First, an intraoperative cholangiogram is performed to delineate the anatomy of the biliary tree and confirm the diagnosis. A liver biopsy is obtained to document degree of liver damage. Next, the fibrotic common bile duct is dissected from the hepatoduodenal ligament up to the level of the porta hepatis and excised. An approximately 20 cm limb of jejunum is brought up in a retrocolic fashion and a Roux-en-Y hepaticojejunostomy is created.

2.4.5 Outcomes

Successful, long term establishment of bile flow correlates with earlier surgical intervention. Infants aged <60 days at time of surgery have best results. Approximately one-third of children undergoing portoenterostomy have a 10-year or greater survival, while the rest will ultimately succumb to liver failure and require transplant. Other indications for liver transplant include presence of intrahepatic atresia, fat soluble vitamin deficiencies causing failure to thrive and variceal bleeding secondary to portal hypertension. 5-year survival following liver transplant ranges from 75 to 95% [2].

Apart from progressive liver failure, cholangitis is another major post-operative complication occurring in as much as 50% of patients who undergo portoenterostomy [2]. Decreased bile flow indicated by elevated total bilirubin in the setting of fever and leukocytosis is essentially diagnostic of cholangitis until proven otherwise. It is managed with IV antibiotics and fluid resuscitation.

2.5 Choledochal cysts

2.5.1 Embryology

Etiology is unknown. Aberrant pancreaticobiliary junction near the duodenal wall has been suggested [3].

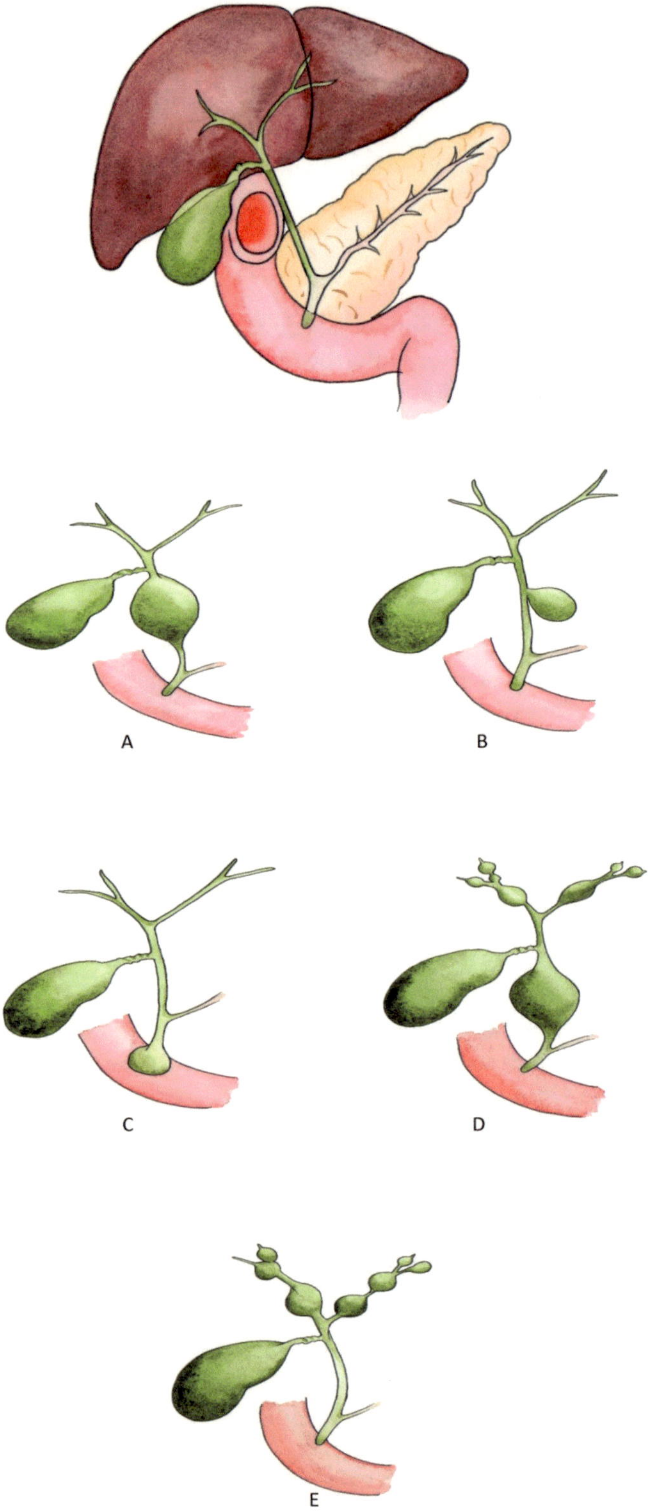

Figure 3.
Normal anatomy of the hepatobiliary tree and its relationship to the pancreas and duodenum. (A) Choledocal cyst type 1: fusiform dilation of the extrahepatic duct common bile duct. (B) Choledocal cyst type 2: isolated diverticulum off the common bile duct. (C) Choledocal cyst type 3: supraduodenal choledococele. (D) Choledocal cyst type 4: cystic dilation of intra- and extra-hepatic bile ducts. (E) Choledocal cyst type 5, dilation of intra-hepatic ducts only.

2.5.2 Clinical presentation

Infants present with symptoms of biliary obstruction: progressive jaundice, dark urine, light colored stools. A tender abdominal mass may be palpated in the right upper quadrant. Laboratory values will be consistent with biliary obstruction and demonstrate elevated direct bilirubin and alkaline phosphatase. Patients may also present with cholangitis or pancreatitis.

2.5.3 Diagnosis

While abdominal ultrasound and hepatobiliary 99-Tc IDA scan are useful, MRCP best delineates the anatomy of the biliary tree and is the diagnostic test of choice. There are five types (**Figure 3**). Type 1 is the most common and presents as saccular or fusiform dilation of the common bile duct (CBD; **Figure 3A**). Intrahepatic ducts are normal. Type 2 is an isolated CBD diverticulum (**Figure 3B**). Type 3 is a choledochocele, in which there is cystic dilation of the supra-duodenal CBD, prior to its junction with the pancreatic duct (**Figure 3C**). In type 4 disease, intra- and extra-hepatic bile ducts are dilated whereas in type 5 disease only intra-hepatic ducts are dilated (**Figures 3D, E**).

2.5.4 Surgical management

Given the risk of cholangiocarcinoma, highest in types I and IV, surgical intervention is indicated at the time of diagnosis of any type of choledochal cyst. The approach depends on type of lesion. For type 1 cysts, primary cyst excision with cholecystectomy and roux-en-Y hepaticojejunostomy reconstruction is procedure of choice. Type 2 disease is managed by simple diverticulectomy. Type 3 is managed by transduodenal cyst excision or marsupialization and sphincteroplasty. Types 4 and 5 may be treated by anatomic hepatic resection based on the extent and location of disease, however, liver transplantation is ultimately required in most cases.

2.5.5 Outcomes

Excision of choledocal cysts result in excellent long-term outcomes with few major complications. Biliary tract malignancy, the most feared complication, may occur with incomplete excision. Cholangitis, stricture formation and choledocolithiasis are lesser significant complications that are managed medically and endoscopically, respectively.

3. Midgut disorders

3.1 Small intestine atresia

3.1.1 Embryology

Midgut development begins around the fifth week of gestation. The midgut starts as a vertical tube and has two connections: a ventral connection to the yolk sac via the omphalomesenteric (vitelline) duct and a dorsal attachment to the posterior abdominal wall, the mesentery [4–6]. The dorsal mesentery is the conduit for the superior mesenteric artery (SMA), which buds from the aorta, and delivers blood to the midgut. The lengthening gut tube outgrows the confines of the abdominal cavity and consequently herniates into the umbilical cord. As it elongates, it rotates

90° in a clockwise direction relative to the embryo (counterclockwise if visualized from the front). The midgut tube continues to grow extra-abdominally during gestational weeks 6–10. Around week 10, it retracts back into the abdominal cavity, rotating another 180° while doing so. Final intra-abdominal growth and fixation ensue, placing the cecum in the right lower quadrant and the duodeno-jejunal junction to the left of the upper midline, inferior to the SMA. The mesentery broadens, fanning out from its root in the posterior abdominal wall, to support the blood vessels and lymphatics that serve the jejunum, ileum, cecum/appendix, ascending colon and proximal 2/3 of the transverse colon. It is believed that ischemic events during this period cause jejunoileal atresia.

3.1.2 Clinical presentation

Atresia causes a structural obstruction that prevents passage of meconium in the first 24–48 h of life and results in bilious emesis. On physical exam, the abdomen will be distended.

Jejunoileal atresia is classified into four types (**Figures 4A–E**). Type 1 is an intraluminal web with intact mesentery (**Figure 4A**). The seromuscular layers of bowel remain in continuity. Type 2 also has an intact mesentery, but the two ends of bowel are disconnected by a fibrous cord (**Figure 4B**). Type 3a has a small v-shaped mesenteric defect that separates two blind ends of bowel (**Figure 4C**). In type 3b disease, known as an "apple-peel" or "Christmas-tree" deformity, a large mesenteric defect separates the proximal and distal ends of bowel. The proximal pouch is very dilated, and the distal collapsed bowel is supplied by a small vessel around which it repeatedly winds (**Figure 4D**). Type 4 consists of numerous blind ended segments of bowel with discontinuous mesentery, appearing as a "string of sausages" (**Figure 4E**).

3.1.3 Diagnosis

Abdominal x-ray will reveal dilated portions of bowel proximal to the site of obstruction with collapsed loops and paucity of air in the distal bowel. Contrast enema will demonstrate an abrupt transition from the filling to non-filling segments of small bowel and the colon will be appear small, <1 cm diameter, due to lack of use. In all cases of bilious emesis, an UGI series is warranted to rule out malrotation, a surgical emergency. UGI will reveal contrast filling in the stomach and proximal bowel, with abrupt cessation of contrast filling at the point of atresia.

3.1.4 Surgical management

Initial management begins with insertion of an oro- or nasogastric tube for bowel decompression and fluid resuscitation. Resection of atretic segments with end-to-end anastomoses is the procedure of choice; however, this can prove quite difficult in cases where ends of bowel are greatly mismatched in diameter. In such circumstances, the anastomosis is created in a fashion similar to duodenoduodenostomy in which the smaller end of bowel is incised longitudinally along its anti-mesenteric border to fit the end of the larger caliber bowel. Prior to completing the anastomosis, the entire length of the bowel must be inspected to ensure there are no intraluminal webs or fenestrations that may cause persistent obstruction. The goal is to resect all defunct bowel segments while maintaining enough length to ensure adequate resorptive capacity. If the ileocecal valve is spared, enteral nutrition can be tolerated with as little as 15–20 cm of small bowel. Otherwise, a length

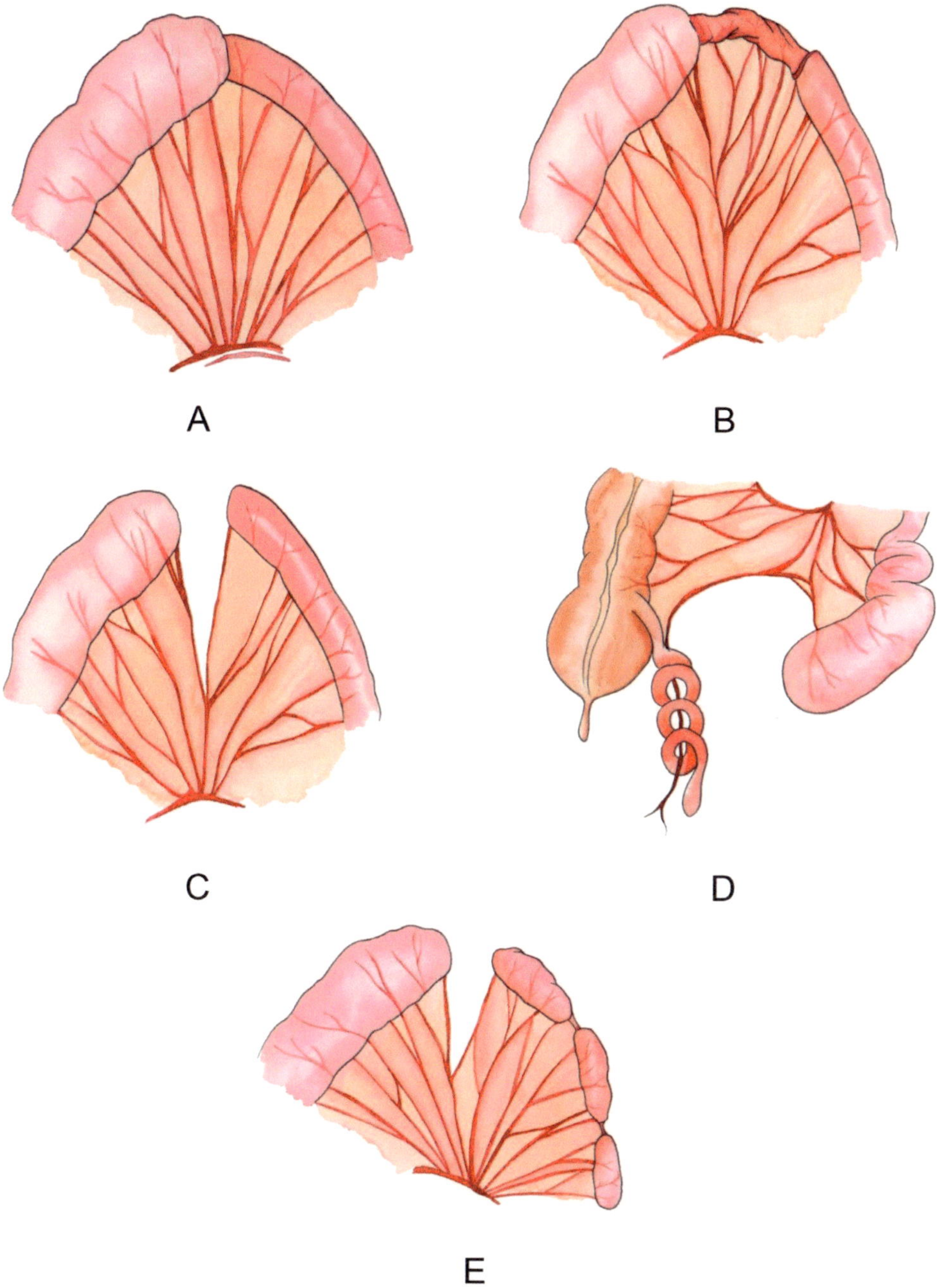

Figure 4.
(A) Type 1 jejunoileal atresia. (B) Type 2 jejunoileal atresia. (C) Type 3a jejunoileal atresia. (D) Type 3b jejunoileal atresia. (E) Type 4 jejunoileal atresia.

of approximately 40 cm is required [4]. Mesenteric defects are closed, taking care not to disrupt the feeding blood vessels.

3.1.5 Outcomes

Intestinal dysmotility, even in infants that have adequate remaining bowel length, may occur for many weeks following repair. Infants with short bowel syndrome, those with less than 40 cm, often require long term parenteral nutrition, which itself carries risks of sepsis and liver damage. Nonetheless, overall mortality is low and related to co-morbidities, such as low birth weight and/or cardiac defects.

3.2 Malrotation and midgut volvulus

3.2.1 Embryology

As described above, normal 270° rotation and fixation of the midgut fails to occur [4–7]. This lack of rotation positions the duodenum and small bowel to the right of the midline and the large bowel to the left. The cecum remains anterior to the duodenum and is tethered to the abdominal wall by lateral peritoneal attachments. These lateral peritoneal attachments, known as Ladd's bands, compress the duodenum, thereby causing obstruction and resultant bilious emesis. The root of the mesentery is narrowed and may potentially act as fulcrum around which the bowel can twist ("volvulize"), thereby kinking the SMA and causing ischemia (**Figure 5**).

3.2.2 Clinical presentation

Acute malrotation with midgut volvulus presents with feeding intolerance and bilious emesis, usually around the first week of life. Abdominal rigidity, overlying erythema are signs of peritonitis and indicate ischemic bowel. Abdominal distention will not be present given the very proximal nature of pathology. As feeding intolerance and bilious emesis are symptoms of multiple pathologies, a high index of suspicion is required to make this diagnosis.

3.2.3 Diagnosis

An abdominal X-ray is typically first obtained, though rarely helpful in establishing the diagnosis. Any concern for malrotation mandates a prompt UGI. A normal study will reveal contrast exiting the pylorus, descending through the second portion of the duodenum and crossing the midline through the third portion of the duodenum into the small bowel. Thus, a normal "C-loop" will be visualized. An abnormal study will demonstrate contrast exiting the pylorus and descending straight down to the right of the midline into the small bowel.

3.2.4 Surgical management

Once the diagnosis of acute malrotation is made, the patient is taken emergently to the operating room for detorsion and evaluation of bowel viability. Fluid resuscitation, insertion of oro- or nasogastric tube for decompression and administration of intravenous antibiotics have ideally been implemented prior to surgical intervention. The bowel is eviscerated and detorsed in a counterclockwise direction, fanning out its mesentery. Ladd's bands are incised to release the obstruction. Any frankly necrotic appearing bowel is resected, while dusky bowel can be re-evaluated and usually salvaged in a second look operation 24–48 h later. Ends of healthy, viable bowel can be anastomosed, otherwise stomas are placed. A prophylactic appendectomy is performed to eliminate the possibility of appendicitis in the future. If a second look operation is required, the abdomen is left open and covered with a temporary sterile dressing; if not, it is closed.

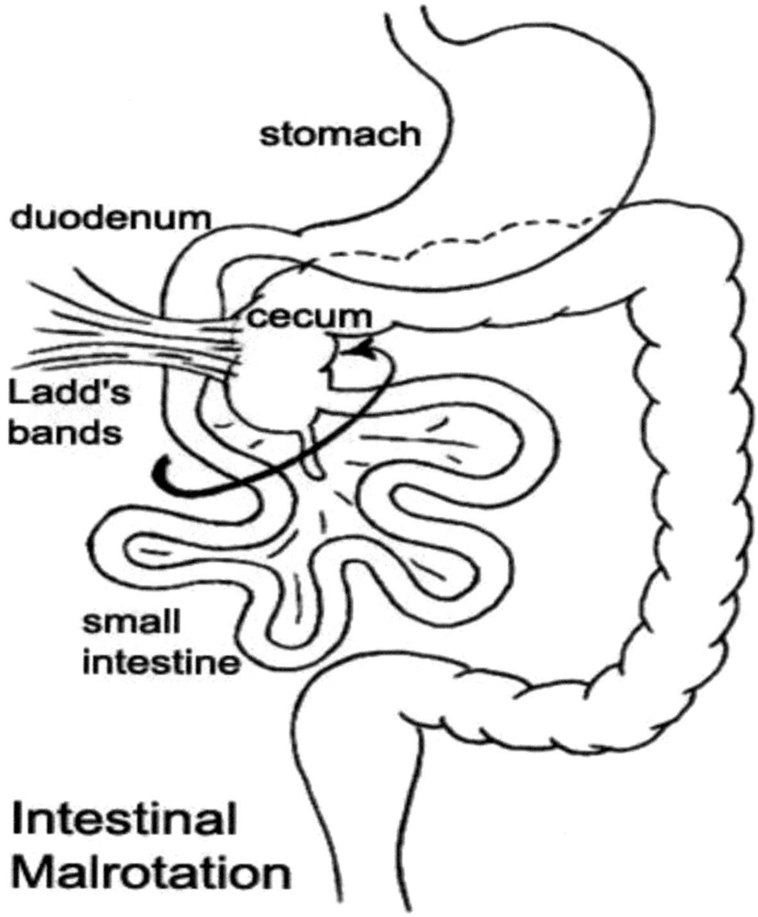

Figure 5.
Intestinal malrotation showing abnormal position of cecum and Ladd's bands

3.2.5 Outcomes

Without significant intestinal necrosis requiring resection, outcomes following correction of malrotation are quite favorable. Infants grow normally and do not have any major adverse sequelae. Rarely, adhesive small bowel obstruction may occur years later, however any operation carries this risk.

3.3 Meckel's diverticulum

3.3.1 Embryology

This condition occurs as a result of the failure of the omphalomesenteric (vitelline) duct to completely involute between weeks 5–7 of gestation (**Figure 6**).

3.3.2 Clinical presentation

Meckel's diverticulum is the most common congenital GIT malformation and the most common cause of painless lower intestinal bleeding in children. It usually presents by the age of 2 years, but presentation can be delayed into the teenage years. There is a male predominance. The bleeding is typically brisk and bright red. Laboratory values will demonstrate anemia. A fibrous cord connecting the diverticulum to the abdominal wall may be present and can act as a point around which bowel can obstruct, twist or intussuscept. In such cases, the child will present with abdominal pain and distention, inability to pass flatus or move their bowels.

3.3.3 Diagnosis

Technetium-99 pertechnate scintigraphy ("Meckel's scan") localizes the bleeding ulcer. The diverticulum is typically found within 2 feet proximal to the ileocecal valve, on the anti-mesenteric side of the ileum and contains heterotopic mucosa, usually that of gastric or pancreatic in origin. Ulceration and bleeding occur secondary to acid secretion from the heterotopic mucosa. It is a true diverticulum involving all four layers of the bowel.

3.3.4 Surgical management

If bleeding is the presenting symptom, ileal resection with primary anastomosis is the procedure of choice. Segmental resection is also indicated in cases complicated by diverticulitis, perforation, obstruction, volvulus or if the base of the diverticulum is very wide. Simple diverticulectomy may be performed if the neck of the diverticulum is narrow, or if diverticulitis does not involve the base.

3.3.5 Outcomes

Resection of Meckel's diverticulum has an excellent prognosis without major long term post-operative complications.

3.4 Omphalocele and gastroschisis

3.4.1 Embryology

These are congenital defects of the abdominal wall, not of the gastrointestinal tract itself, but are discussed because they are associated with malrotation (**Figure 6**).

3.4.2 Clinical presentation

Numerous physical characteristics differentiate omphalocele from gastroschisis. The abdominal wall defect in omphalocele is midline, versus to the right of the umbilicus in gastroschisis. Defects tend to be smaller in gastroschisis, typically ≤3 cm. In comparison, omphaloceles can vary widely in diameter, ranging in size from 2 to 15 cm. Larger defects allow for herniation of more organs, namely the liver and spleen. This rarely, if at all, occurs in gastroschisis. Herniated contents are covered by an amniotic sac in omphalocele but not in gastroschisis. Exposure of the bowel to amniotic fluid during gestation causes the bowel to become thickened and the mesentery fibrotic whereas bowel is normal in omphalocele since it is protected by the overlying sac. Lastly, omphalocele has a higher association with chromosomal abnormalities and other congenital anomalies compared gastroschisis. Intestinal atresia may be seen in gastroschisis.

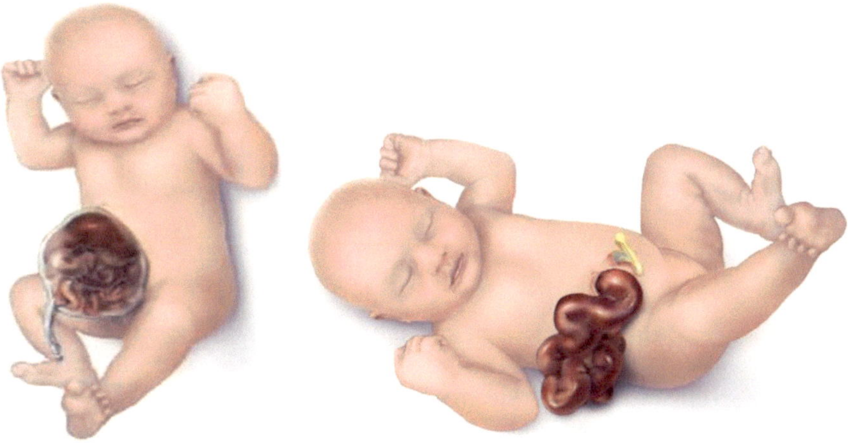

Figure 6.
Omphalocele (left) and gastroschisis (right). The herniated intestine is covered with a sac with umbilical cord attached to it in omphalocele, while the intestinal loops in gastroschisis herniate through a defect on the right side of umbilicus and are not covered.

3.4.3 Diagnosis

These defects may be appreciated on pre-natal ultrasound and are therefore expected upon delivery. Chest radiography, echocardiogram and renal ultrasound are performed to rule out associated anomalies in the case of omphalocele, as is karyotyping though this may have been performed prenatally.

3.4.4 Surgical management

Exposure of intestinal contents to the environment can result in significant insensible losses. Initial management aims to maintain adequate volume status and body temperature. The infant is placed under a warmer, fluid resuscitation commenced, and urinary catheter inserted to strictly monitor volume status. Oro- or naso-gastric tube is placed for bowel decompression. Intestinal contents are wrapped in a moist, sterile plastic dressing to prevent evaporative losses. In the case of omphalocele, care must be taken to prevent rupture of the protective sac. The goals of operation are to return the herniated contents into the abdominal cavity and close the defect. If this is unable to be accomplished either because the infant is too unstable to be taken to the operating room or because there is high risk of abdominal compartment syndrome, a silo can be sutured in place over the herniated viscera and contents gradually reduced. Daily manual reduction can be performed bedside, gently as tolerated, with complete reduction usually achieved over 3–7 days. The resultant ventral hernia is repaired once all viscera have been reduced and the infant deemed fit to tolerate general anesthesia.

3.4.5 Outcomes

Given the protective nature of the overlying sac in omphalocele, infants typically have normal bowel function following reduction and abdominal wall repair. Long term complications are related to concomitant congenital defects. In contrast, patients with gastroschisis, especially if they also have intestinal atresia, are subject to dysmotility, malabsorption and are at increased risk of developing necrotizing enterocolitis. These infants often require long term parenteral nutrition following surgical correction.

4. Hindgut disorders

4.1 Hirschprung's disease

4.1.1 Embryology

Aganglionosis of the myenteric plexus due to failure of neural crest cell migration during weeks 6–12 of embryonic development. Most often occurs in the rectum though any portion and, rarely, the entire bowel can be affected. The myenteric plexus lies in between the outer longitudinal and inner circular muscle layers of the colon and is responsible for peristalsis.

4.1.2 Clinical presentation

Aganglionosis results in a functional obstruction manifesting as failure to pass meconium within first 24 h of life. Abdominal distention may be present. Rectal stimulation causes explosive passage of air and stool. Because disease is distal, infant will likely be able to tolerate oral intake though may have intermittent episodes of bilious emesis. Less severe disease may not manifest until later in childhood, up to 2–3 years of age, with chronic constipation. There is an association with trisomy 21. Therefore, work up includes echocardiogram to rule out concomitant cardiac defects.

4.1.3 Diagnosis

Gold standard is suction rectal biopsy, which demonstrates aganglionosis of the myenteric plexus. Biopsy should be obtained 1–1.5 cm proximal from the dentate line to ensure rectal specimen is obtained. Pathology will reveal unmyelinated nerve fibers with hypertrophied endings that stain darkly with acetylcholinesterase. Abdominal X-ray shows dilated proximal bowel with collapsed distal colon. Contrast enema is helpful in distinguishing transition zone between affected and normal colon however, gross anatomic distinction does not always correlate with histopathology [8].

4.1.4 Surgical management

Although various operative methods have been described, the fundamental principle of each procedure is the same: to establish continuity between the normal, ganglionic segments of bowel. In the past, multi-stage operations beginning with decompressive colostomy followed by definitive repair was common. Nowadays, single-stage laparoscopic approach is preferred. Regardless of procedure, however, intra-operative frozen section must be performed to confirm the presence of normal ganglionic colon prior to anastomosis, otherwise dysfunction will continue post-operatively.

4.1.4.1 Swenson-original procedure

The rectum/aganglionic segment is dissected circumferentially, everted through the anus and resected. Normal colon is pulled down and a low end-to-end colorectal anastomosis is created.

4.1.4.2 Duhamel procedure

The aganglionic portion of bowel is bypassed and a posterior end to side anastomosis is created between the innervated segments of colon and distal rectum. The rectum is stapled at the proximal margin of disease. An incision is made in the distal posterior wall of the rectal stump approximately 1 cm superior to the dentate line. The innervated colon is pulled down through the presacral space and then anastomosed in an end-to-side fashion to the distal posterior rectal wall. The defunct rectal stump is left in place.

4.1.4.3 Soave procedure

Circumferential endorectal dissection of rectal mucosa and submucosa, followed by evagination of these layers through the anus for resection. A rectal muscular channel remains, and innervated colon is intussuscepted through the remaining rectal muscular channel. A colorectal anastomosis is performed at the distal end of the muscular channel [9].

4.1.5 Outcomes

No single procedure has been shown to be superior to other in terms of long-term outcomes, and up to 90% patients will have relatively normal bowel function following repair. Although results tend to be quite favorable, one significant cause of significant morbidity and mortality is Hirschsprung's enterocolitis. While the exact etiology of this entity is unknown, bacterial overgrowth and translocation appear to be implicated. Patients present with fever, abdominal distention and diarrhea. Management consists of fluid resuscitation, IV antibiotics and rectal irrigation. Refractory cases require surgical decompression with a proximal ostomy. Other complications such as anastomotic leak, stricture, abscess, wound infection and obstruction occur in up to 10% cases [1].

4.2 Anorectal malformation/imperforate anus

4.2.1 Embryology

During the 5th week of gestation, the midline urorectal septum descends in a caudal direction toward the cloaca and divides into ventral and dorsal portions. The ventral bud becomes the urogenital sinus, which develops into the urethra and bladder. The dorsal bud becomes the rectum and anal membrane. The anal membrane involutes around week 8, thereby forming the anus. Dysgenesis can occur at any time point, allowing for variability in clinical presentation.

An anatomical distinction based on the pathology's relation to the levator ani muscle complex was first described by Pena. The levator ani complex supports the pelvic floor and is composed of three striated muscles: the puborectalis, the pubococcygeus and the iliococcygeus. The puborectalis encircles the base of the rectum, helps to form the external anal sphincter and thereby plays an integral role in regulating defecation. Anorectal dysgenesis above the levator ani muscles is considered a "high" lesion. Conversely, lesions inferior to the levator ani complex are termed "low" malformations. Generally speaking, higher malformations tend to cause more severe issues with controlling defecation as the neuromuscular development between the levator ani complex and growing recto-anus is compromised to a greater degree.

4.2.2 Clinical presentation

Failure to pass meconium in the first 24–48 h of life. Physical exam will reveal abdominal distention and absence of anus. A subtle opening in the perineum through which small amounts of meconium pass may be present and indicates an anoperineal fistula in the setting of a low imperforate anus. This is the most common pathology seen. In females, low lesions may also be associated with a rectovestibular fistula, and meconium may be expressed through the vagina. Elimination of meconium during urination indicates rectourethral or rectovesicular fistula and a high rectal pouch.

4.2.3 Diagnosis

Diagnosis is made upon physical examination of the perineum. Historically, an invertogram was performed to evaluate the length of atresia. In this study, a radiopaque marker is placed on the infant's bottom, where the anus would normally be located, and the infant is placed in a head down position to allow air to ascend at the most inferior point in the rectum. Lateral films of the pelvis are then obtained. The distance between the marker and distal rectum indicate the level of pathology—high vs. low. Now, ultrasound is preferred.

Anorectal malformations are part of the VACTERL syndrome and most commonly associated with concomitant genitourinary defects. In addition to a renal ultrasound, a voiding cystourethrogram should be obtained, especially if a recto-urethral/rectovesicular fistula is suspected as this can help delineate the tract. Plains films of the chest, limbs and spine as well as an echocardiogram help identify the presence of other anomalies. Any other life-threatening co-morbidities take precedence, and a temporary diverting ostomy can be placed until definitive repair can be safely performed, usually between 8 and 12 months of age.

4.2.4 Surgical management

Posterior sagittal anorectoplasty (PSARP) is the surgical procedure performed. The infant is placed in a prone jack-knife position. If a perineal fistula is present, an incision is made around the fistula and carried posteriorly toward the coccyx. If no perineal fistula is present, the incision starts inferior to the coccyx and is carried down to the perineum. It is imperative to remain midline. This is ensured by visualizing striated muscle fibers, which run perpendicular to the incision. If fat is encountered during the dissection, this indicates that the operator has deviated from midline and entered the lateral ischioanal/ischiorectal space. The rectum is identified by its overlying glistening fascia and then freed circumferentially, beginning posteriorly and advancing anteriorly until the fistula is encountered. The fistula is resected. After the fistula is taken down, the anterior rectal wall is freed from its surrounding structures. In females, the anterior rectum lies in close proximity to the posterior vaginal wall and in males, the prostate and bladder. The anterior rectal wall is gently dissected off these structures up to the peritoneal reflection. Complete, circumferential dissection of the rectum will allow for tension-free pull down and anastomosis. The rectum is situated in its anatomic position in the muscle complex. The muscle complex is repaired around the properly positioned rectum and the neoanus is created by suturing mucosa to the perineum.

4.2.5 Outcomes

Long terms outcomes are dependent on the level of pathology—high versus low anorectal dysgenesis—and the extent of neuromuscular development of the levator ani complex and rectum. Almost all children will require some degree of lifestyle modifications to manage fecal incontinence or, conversely, chronic constipation. This is achieved by strict bowel regimens with enemas or cathartics. In more severe cases, a cecostomy or appendicostomy may be required to allow for daily antegrade enemas. Worst case scenarios may necessitate a diverting ostomy.

Acknowledgements

The contribution of Natalia Louise Smith is greatly appreciated for drawing the figures numbered as 1-to-6.

Author details

Richa Verma
Piedmont Healthcare, Statesville, NC, USA

*Address all correspondence to: rverma1024@gmail.com

References

[1] Abdellatif M, Ghozy S, Kamel MG, et al. Association between exposure to macrolides and the development of infantile hypertrophic pyloric stenosis: A systematic review and meta-analysis. European Journal of Pediatrics. 2019;**178**:301-314. DOI: 10.1007/s00431-018-3287-7

[2] Greenfield Lazar J, Mulholland Michael W, editors. Greenfield's Surgery: Scientific Principles and Practice. 5th ed. Philadelphia: Wolters Kluwer Health/Lippincott Williams & Wilkins; 2011

[3] Soares KC, Goldstein SD, Ghaseb MA, et al. Pediatric Surgery International. 2017;**33**:637. DOI: 10.1007/s00383-017-4083-6

[4] Fiser SM. The ABSITE Review: Practice Questions. 2nd ed. Richmond, VA: Hancock Surgical Consultants; 2015

[5] Dimick JB, Upchruch JR, Sonnenday CJ, editors. Clinical Scenarios in Surgery: Decision Making and Operative Technique. Philadelphia: Wolters Kluwer Health/Lippincott Williams & Wilkins; 2012

[6] http://www.embryology.ch/anglais/sdigestive/mitteldarm01.html

[7] https://discovery.lifemapsc.com/library/review-of-medical-embryology/chapter-84-development-of-the-midgut-general-introduction https://www.pedsurglibrary.com/apsa/view/Pediatric-Surgery-NaT/829186/all/Posterior_Sagittal_Anorectoplasty#0

[8] https://coloproctol.org/journal/view.php?number=1558

[9] Morris G, Kennedy A, Cochran W. Current Gastroenterology Reports. 2016;**18**:16. DOI: 10.1007/s11894-016-0490-4

Chapter 6

Common Congenital Neural Tube Anomalies: Epidemiology, Classification, Management and Outcome

Mohammad Hossein Khosravi and Bita Najafian

Abstract

The prevalence of Congenital central nervous system (CNS) anomalies, including those of the brain and spinal cord, is 3 to 6% in stillbirth and 0.14 to 0.16% in live births. Holoprosencephaly, spina bifida, anencephaly, and encephalocele are major neural tube defects (NTD) encountered in clinical practice. Proper management and diagnosis of these conditions mandate a good understanding of their etiology and classification. Research is being conducted to investigate the etiopathogenesis and treatment of these anomalies. In this chapter, we have reviewed the clinical and pathological aspects of the major NTDs and the latest principles of their management.

Keywords: the central nervous system, congenital anomalies, Fetal CNS anomalies, Neural Tube Defects

1. Introduction

Until recently, central nervous system (CNS) malformations were the second most common congenital abnormalities after congenital cardiac defects [1, 2]. Recent reports have documented CNS malformations to be the most common anomalies among all systems, with a prevalence of 3 to 6% in stillbirth and 0.14 to 0.16% in live births [3]. There is limited information about the precise etiology of congenital CNS anomalies, and most of the cases are idiopathic. It is speculated that a combination of genetic and environmental factors plays a major role in the pathogenesis of these defects [4]. Management and diagnosis of these conditions are challenging and require a proper understanding of their etiology and categories.

CNS anomalies (CNSA) include those of the spinal cord (such as meningocele, myelomeningocele, and encephalocele) and brain (including growth disorders of the cerebrum, cerebellum, and brain stem) [5]. CNSA may be associated with other anomalies pertaining to other systems as well, such as those of the heart [6]. These malformations need complex surgeries along with long-term intensive care and impose a significant financial impact on the families and healthcare system. In this chapter, we review the classification, epidemiology, and the newest modalities of treatment of congenital CNS anomalies with respect to NTDs.

2. Incidence

Despite remarkable developments in diagnostic technologies and therapeutic modalities, the epidemiology of congenital CNS anomalies has not changed significantly. The prevalence of CNSA varies widely according to geographic regions and socioeconomic situations and is reported to be between 1 and 10 in every 1000 live births [7, 8]. The incidences of anencephaly and spina bifida per 10000 births range from 0.7 in central France to 0.9 in Canada, 7.7 in the United Arab Emirates, and 11.7 in South America [9]. A recently published systematic review and meta-analysis, which included 6558 infants, has reported a prevalence of occult spinal dysraphism (OSD) with cutaneous stigmata to be 2.8% [10].

3. Classification

In general, NTDs are classified into open and closed defects [11].

3.1 Open NTDs

3.1.1 Craniorachischisis

This is the most severe presentation of NTDs and involves both the spinal and cranial parts of the neural tube (**Figure 1**) [12]. Craniorachischisis is a combination of anencephaly with a contiguous bony defect of the spine, both without the neural tissue's meningeal cover.

3.1.2 Iniencephaly

It is a rare severe defect of the occipital bone, with cervical spina bifida and retroflexion of the head on the cervical spine. An occipital encephalocele may be present. Like anencephaly, there is a strong female preponderance (**Figure 2**).

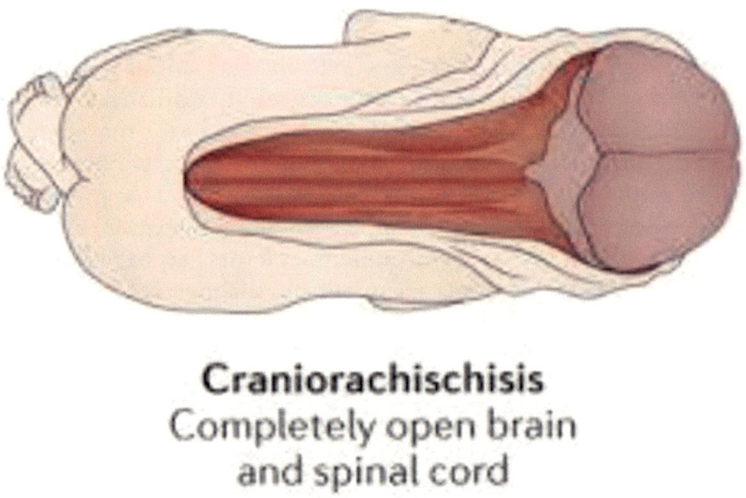

Figure 1.
Diagram of Craniorachischisis.

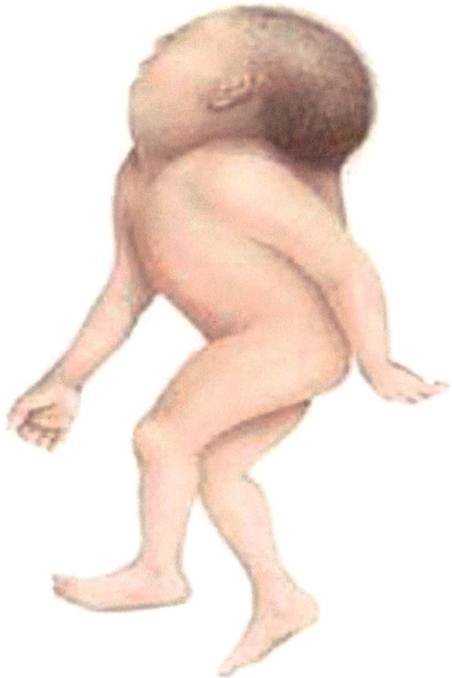

Figure 2.
Diagram of iniencephaly.

3.1.3 Anencephaly

In this defect, the cranial portion of the neural tube fails to close, resulting in exencephaly. This ends up in the neural tissue getting a destructive exposure to an intra-amniotic environment which turns the exencephaly into anencephaly (**Figure 3**) [12].

3.1.4 Myelomeningocele

In this defect, the posterior part of the spinal portion of the neural tube fails to close; This defect occurs most commonly in the lumbar region. A bony defect in the vertebral arch provides the condition for the meningeal sac to herniate (**Figure 4**). Myelocele is a similar condition that involves the spinal cord without protrusion of the meningeal sac.

3.2 Closed NTDs

3.2.1 Encephalocele

It is defined as a sac-like protrusion of the brain accompanied with or without meninges through an opening in the skull (**Figure 5**). According to the type of involved tissues, encephaloceles are classified as meningocele (herniation of meninges), encephalomeningocele (herniation of both meninges and brain), and encephalomeningocystocele (herniation of meninges, brain, and ventricle).

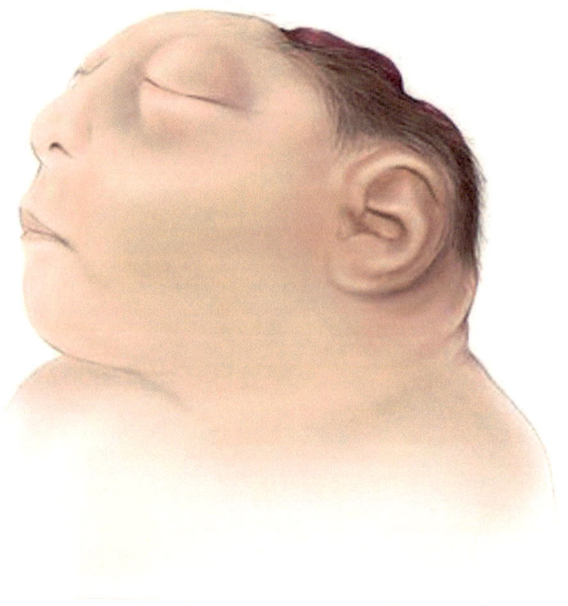

Figure 3.
Diagram showing anencephaly.

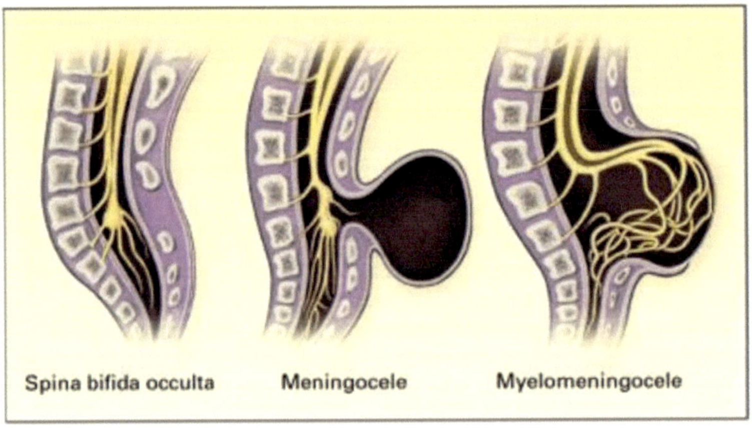

Figure 4.
Diagram showing spina bifida occulta, memingocele and myelomeningocele.

3.2.2 Meningocele

Meningocele is the protrusion of meninges through the vertebral arch defect without the spinal cord (**Figure 4**) [13]. This defect is macroscopically similar to myelomeningocele with differences in the contents of the herniated sac.

3.2.3 Spina bifida Occulta

Abnormal development of the embryonic tail bud results in a wide range of spinal cord abnormalities grouped as spina bifida occulta (**Figure 2**). It is generally accompanied by other skeletal defects such as sacral agenesis. The anomaly mostly involves sacral and lower lumbar vertebrae.

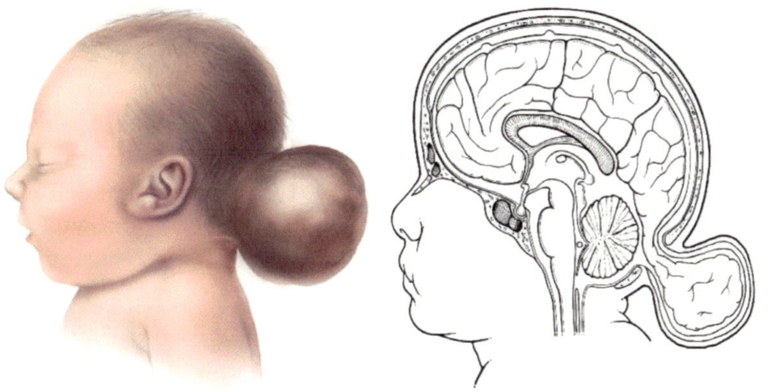

Figure 5.
Diagrams of encephalocele.

4. Etiology and embryology

Despite recent progress in epidemiologic and clinical research, the exact etiology of NTDs has remained undetermined. It is in common agreement that the interactions between genetic and environmental factors are the possible etiopathogenic factors [14, 15]. More than 70% of cases of NTDs are found to have a genetic etiology [16].

Some of the potential non-genetic environmental risk factors for NTD-affected pregnancy are poor socioeconomic status, maternal hyperthermia; maternal exposure to high doses of irradiation, certain chemicals, and drugs; cigarette smoking; maternal metabolic diseases, and advanced parental age, including those of both mother and father. Pregnancies associated with a fetus with NTD have higher chances of going into preterm labor and being delivered prematurely [5, 17, 18]. In 2008, a large study in California reported that mothers who do not graduate from high school or live in neighborhoods under poor socioeconomic conditions have a greater risk of delivering an NTD-affected child [19]. Brough et al. elaborated in their study that the mothers with higher socioeconomic and educational levels are more likely to consume folic acid during preconception and early gestational age when the neural tube is developing [20], and this might have contributed to the findings of the California study. A meta-analysis regarding the effect of maternal age on the risk of NTD reported that mothers older than 40 and younger than 19 years of age had increased risks of NTD-affected pregnancies [21], the chances being higher for spinal Bifida but not for anencephaly. Studies have assessed the role of parental occupational exposures in the development of NTDs. Brender et al. found self-reported multiple pesticide exposure to be a risk factor for fetal NTD [22]. The role of paternal exposures to hazardous materials in increasing the risk of NTDs in offsprings was emphasized when studies showed that fathers who work as a cook, gardener, janitor, and cleaner have higher chances of getting a child with spina bifida, as these professions have a higher likelihood of exposure to hazardous chemicals [23]. The occupational exposure of fathers to metal-working oil mists and hydrocarbons do not show any association with NTD risk [24]. Caffeine has been investigated as another risk factor for NTDs. Past studies have shown that higher Caffeine consumption during the year before pregnancy increases the risk of spina bifida [25]. The use of antimicrobial medications during the preconception period and first trimester of pregnancy are found to be associated with a higher risk of anencephaly [26].

4.1 Genetics of congenital CNS anomalies

In addition to the potential environmental risk factors as outlined above, congenital CNSA may be a consequence of genetic disorders. NTD, including encephalocele, spina bifida, and exencephaly, have become less prevalent since the widespread consumption of folic acid by pregnant women. So the etiology has shifted toward mutations in folate-responsive or folate-dependent pathways [27]. Knowing the underlying genetic disorders for congenital CNS anomalies helps in counseling about the existing pregnancy, as well as the risk of recurrence in future ones. Sophisticated genetic investigations are available to detect chromosomal anomalies in the cases of NTD.

While a low rate of karyotype abnormality has been reported in isolated ventriculomegaly (0 to 3.8%) [28], Dandy-Walker malformation has been associated with 50% of aneuploidies if associated with other anomalies [29]. In a nearly similar pattern, isolated holoprosencephaly is not accompanied by any significant genetic anomaly; however, 25 to 50% of cases have been reported to have aneuploidies if associated with other organ anomalies. Holoprosencephaly is detected in 70% of Trisomy 13 cases. In the cases with NTDs, Trisomy 13 and 18 are the most commonly reported aneuploidies [30]. Studies have reported notable connections between deletions on the long arm of the 13th chromosome and CNS anomalies [31].

Genes participating in folate metabolism have been studied for the pathogenesis of NTDs. The C677T and A1298C polymorphisms of methylenetetrahydrofolate reductase (MTHFR), encode a key enzyme of folate metabolism responsible for homocysteine remethylation [16]. These polymorphisms are associated with a 1.8fold increase in the risk of NTDs. More than 200 genetic models of NTD have been described in mice that can affect the open NTD phenotypes, such as anencephaly, open spina bifida and craniorachischisis. Their roles in human NTDs are understudy. Some inbred strain variation in the penetrance and expressivity of NTD phenotypes in mice have suggested the roles of modifier gene function. The Cecr2 mutation that causes exencephaly in mice is strongly affected in its expression by one or more modifier genes on Chromosome 19. Strain differences have also been described for non-genetic causes of NTD, including hypoglycemia, hyperthermia, valproic acid, and cytochalasins.

4.2 Embryologic formation of neural tube

Formation of the brain and spinal cord begins with the development of the neural tube through the embryonic process of neurulation. The neural tube is the origin of the brain and spinal cord. The process of neurulation, has two separate phases in mammalian embryos, termed primary and secondary neurulation [32].

Primary neurulation occurs in the third and fourth weeks of development, during which the flat layer of ectodermal cells is transformed into a hollow tube. On the 18th day of fertilization, the neural plate is formed by a thickening of the embryonal midline dorsal ectoderm. The neural plate develops from the cranial end of the embryo moving toward the caudal end. *This sentence is deleted* ---Then, the edges of the neural plate move upward to form the neural fold---. On 19th day, the border of the neural plate becomes elevated and folds longitudinally from the head to the tail, which results in the formation of a neural groove. By the 23rd day, the folds get merged and make the neural tube which is open at both ends. "Closure," a process in which both open ends of the neural tube (neuropores) are closed, occurs on the 26th and 28th day of gestational age in rostral and caudal ends, respectively. This neural tube closure is initiated at the hindbrain/cervical boundary (Closure 1).

Fusion extends into the hindbrain and along the spinal region, leading to several closure sites appearing at the midbrain/forebrain boundary (Closure 2) and at the rostral extremity of the forebrain (Closure 3). The progression of fusion ('zippering') continues along the spine, ending in the last closure at the posterior neuropore at the level of the second sacral segment. This process of neural folding is called 'primary' neurulation.

Stem cell proliferation is the main mechanism involved in secondary neurulation, which is limited to the tailbud. During this process, a rod-like condensation is formed, which subsequently becomes cavitated. Cavitation transforms the rod into a tube which remains in continuation with the tube constructed as the result of primary neurulation. Tail bud develops in tailed mammalians, and as the anatomy of humans is tailless, secondary neurulation does not seem to be involved in the formation of neural tube defects [8]. Instead, the lateral sclerotome cells derived from the multipotential tail bud cell population organize themselves around the secondary neural tube to form the sacral and coccygeal vertebrae. Subsequently, the caudal-most neural tube degenerates via apoptosis.

Primary neurulation is essential for the formation of the brain and spinal cord. Failure of Closure 1 leads to the most severe NTD, called craniorachischisis, which comprises an open neural tube encompassing the midbrain, hindbrain, and spinal region. In the presence of normal completion of closure one incomplete closure of the cranial neural tube leads to anencephaly with may have the defect confined to the midbrain (meroanencephaly) or extending into the hindbrain (holoanencephaly). Failure of Closure 3 is uncommon and presents with abnormal face with anencephaly. In the spinal region, failure of final closure at the posterior neuropore yields an open spina bifida (myelomeningocele). In this anomaly, the upper limit of SB depends upon the timing of the arrest of the progression of zippering and, as such, may end up at varying axial levels.

A hypothesis forwarded by Morgagni believes that the increased intraventricular pressure caused by over-production of cerebrospinal fluid (CSF), leads to the reopening of an already closed neural tube [33–35].

5. Diagnosis

5.1 Laboratory-based diagnosis of NTDs

Maternal serum alpha-fetoprotein blood levels are used for the screening of CNS anomalies in the fetus, in addition to magnetic resonance imaging or ultrasonography [36, 37].

5.2 Imaging-based diagnosis of NTDs

Fetal magnetic resonance imaging (MRI) was first reported in 1983 [38]. In the late 1990s, fast-sequence MRI was introduced (which eliminated the need for maternal sedation), and fetal MRI was preferred by clinicians [39]. Many studies have reported MRI to be a more accurate technique for diagnosing fetal CNS anomalies compared to ultrasound [40, 41].

In 2014, a systematic review was conducted by Rossi et al., which included 13 original articles and 710 fetuses. This report documented that in addition to confirming the US findings in 65.4% of cases, MRI provides additional information (especially about midline anomalies) in 22.1% of cases [37]. Overall, MRI was able to identify CNS anomalies in 18.4% of cases. Ultrasound was more accurate than MRI in 2% of cases. In 30% of cases, the MRI findings of fetal visualization were different from US findings enough to change the management. They reported a false-positive rate of 2.5% in diagnosing conditions like midline anomalies, hemorrhage, and cell-proliferation disorders by MRI. It is suggested that clinicians should combine fetal MRI with 2-or 3-D-US in order to reduce false-positive diagnosis and increase the sensitivity [42].

It is known that the placenta plays a key role in the fetal development and in protecting the fetus against the maternal immune system and pathogens. There are correlations between placental dysfunction and neurodevelopmental injury [43, 44], and placental ischemia and inflammation can damage the developing fetal CNS. Fetal MRI provides the opportunity to accurately assess *invivo* fetal placental and brain function [45]. In intrauterine growth restricted conditions, placentas have decreased volume as well as lower apparent diffusion coefficient (ADC) values [46–48]. Shapira-Zaltsberg et al., in an interesting study in 2017, evaluated the MRI characteristics of the placenta in fetuses with and without CNS anomalies. They concluded that in diffusion-weighted imaging (DWI) of fetal MRI, restricted diffusion in placenta as well as reduced ADC values are accompanied with fetal CNS abnormalities [43].

6. Clinical presentation, management, and outcomes

Clinical presentations are highly dependent on the type, size, and location of abnormalities, varying from no evident symptoms to lifelong disabilities and even death [10].

6.1 Craniorachischisis

This anomaly is lethal and has no cure or surgical management.

6.2 Anencephaly

Being a condition incompatible with survival, anencephaly diagnosed during early pregnancy may result in a legal interruption of pregnancy. The majority of anencephalic newborns die within the first day of birth. Surgical treatment is not indicated [49].

6.3 Myelomeningocele

If the infant is not affected by other serious anomalies or malformations, most cases with myelomeningocele or myelocele survive with a wide range of neurological impairments. The level of the defect is a determining factor of the clinical characteristics. In levels below the spinal lesion, patients may face a variety of motor and sensory deficits, including bladder incontinence or sexual dysfunction [50]. Most of these lesions are managed by surgical intervention, followed by rehabilitation.

6.4 Encephalocele

Content and location of the herniated mass is a predictive marker of prognosis and clinical manifestations. "The more rostral the site, the better the prognosis" [50]. Epilepsy, sensorial or motor neuron dysfunction, or various degrees of developmental deficiencies may occur in cases due to mechanical effects of traction and distortion on the brain stem [51]. Some patients may take benefits from surgical intervention.

6.5 Meningocele

Patients usually have a normal neurological examination with no evident sphincter dysfunction or deformity of the lower extremities. A simple surgical correction is the main treatment method.

6.6 Spina Bifida Occulta

Children with spina bifida have a higher first-year mortality rate in comparison with the general population [14]. Researchers have shown that children with Spina Bifida have increased sedentary behaviors and lower physical activity levels than their healthy counterparts [52]. This increases the risk of various comorbidities such as diabetes, obesity, thrombosis, etc. Most commonly, this malformation is diagnosed later in life as it has no evident neurological manifestations and disabilities. The neurological symptoms may occur when the spinal cord faces damage or traction. In symptomatic patients, neurosurgical intervention is the main therapeutic method.

7. Future direction

An extensive review of the literature shows a dearth of epidemiologic and etiologic studies. More original studies and meta-analyses are needed to understand the genetic and environmental risk factors of NTDs. Although most of the established surgical interventions have positive effects, prevention remains the best strategy in the management of NTDs.

Author details

Mohammad Hossein Khosravi[1,2] and Bita Najafian[3*]

1 Functional Neurosurgery Research Center, Shohada Tajrish Comprehensive Neurosurgical Center of Excellence, Shahid Beheshti University of Medical Sciences, Tehran, Iran

2 Department of Research, Arka Education and Clinical Research Consultants, Tehran, Iran

3 Department of Pediatrics, Faculty of Medicine, Baqiyatallah University of Medical Sciences, Tehran, Iran

*Address all correspondence to: dr.najafian@yahoo.com

References

[1] Icenogle DA, Kaplan AM. A review of congenital neurologic malformations. Clinical pediatrics. 1981;20(9):565-76.

[2] Aydın E, Tanacan A, Büyükeren M, Uçkan H, Yurdakök M, Beksaç MS. Congenital central nervous system anomalies: Ten-year single center experience on a challenging issue in perinatal medicine. Journal of the Turkish German Gynecological Association. 2019;20(3):170.

[3] Ekmekci E, Demirel E, Gencdal S. Conventional Chromosome Analysis of Fetuses with Central Nervous System Anomalies and Associated Anomalies: Is Anything Changed? Medical Sciences. 2018;6(1):10.

[4] Mashuda F, Zuechner A, Chalya PL, Kidenya BR, Manyama M. Pattern and factors associated with congenital anomalies among young infants admitted at Bugando medical centre, Mwanza, Tanzania. BMC research notes. 2014;7(1):1-7.

[5] Kinsman SL, Johnston MV. Kliegman RM, Stanton FB, Jenson HB, Behrman RE. Nelson Text book of pediatrics. 19th ed. Philadelphia: Saunders; 2011. Congenital anomalies of the central nervous system; pp. 1998-2013.

[6] Hadžagić-Ćatibušić F, Maksić H, Užičanin S, Heljić S, Zubčević S, Merhemić Z, et al. Congenital malformations of the central nervous system: clinical approach. Bosnian Journal of Basic Medical Sciences. 2008;8(4):356.

[7] De LN, Medeiros F, Martins V, Sampaio G, Serapiao M, Kastin G, et al. Malformations of the central nervous system: analysis of 157 pediatric autopsies. Arquivos de neuro-psiquiatria. 2000;58(3B):890-6.

[8] Detrait ER, George TM, Etchevers HC, Gilbert JR, Vekemans M, Speer MC. Human neural tube defects: developmental biology, epidemiology, and genetics. Neurotoxicology and teratology. 2005;27(3):515-24.

[9] Rimoin DL, Connor JM, Pyeritz RE, Korf BR. Emery and Rimoin's principles and practice of medical genetics: Churchill Livingstone Elsevier; 2007.

[10] Choi SJ, Yoon HM, Hwang JS, Suh CH, Jung AY, Cho YA, et al. Incidence of occult spinal dysraphism among infants with cutaneous stigmata and proportion managed with neurosurgery: a systematic review and meta-analysis. JAMA network open. 2020;3(7):e207221-e.

[11] Avagliano L, Massa V, George TM, Qureshy S, Bulfamante GP, Finnell RH. Overview on neural tube defects: From development to physical characteristics. Birth defects research. 2019;111(19): 1455-67.

[12] Stewart W. Pathology and genetics: developmental neuropathology. Wiley Online Library; 2005.

[13] McComb JG. A practical clinical classification of spinal neural tube defects. Child's Nervous System. 2015;31(10):1641-57.

[14] Frey L, Hauser WA. Epidemiology of neural tube defects. Epilepsia. 2003;44:4-13.

[15] Padmanabhan R. Etiology, pathogenesis and prevention of neural tube defects. Congenital anomalies. 2006;46(2):55-67.

[16] Copp AJ, Greene ND. Neural tube defects--disorders of neurulation and related embryonic processes. Wiley Interdiscip Rev Dev Biol. 2013 Mar-Apr;2(2):213-27. DOI: 10.1002/wdev.71.

Epub 2012 May 29. PMID: 24009034; PMCID: PMC4023228.

[17] Taksande A, Vilhekar K, Chaturvedi P, Jain M. Congenital malformations at birth in Central India: A rural medical college hospital based data. Indian journal of human genetics. 2010;16(3):159.

[18] Little J, Cardy A, Arslan M, Gilmour M, Mossey P, includes IMc. Smoking and orofacial clefts: a United Kingdom–based case-control study. The Cleft palate-craniofacial journal. 2004;41(4):381-6.

[19] Grewal J, Carmichael SL, Song J, Shaw GM. Neural tube defects: an analysis of neighbourhood-and individual-level socio-economic characteristics. Paediatric and perinatal epidemiology. 2009;23(2):116-24.

[20] Brough L, Rees G, Crawford M, Dorman E. Social and ethnic differences in folic acid use preconception and during early pregnancy in the UK: effect on maternal folate status. Journal of human nutrition and dietetics. 2009;22(2):100-7.

[21] Vieira AR. Maternal age and neural tube defects: evidence for a greater effect in spina bifida than in anencephaly. Revista medica de Chile. 2005;133(1): 62-70.

[22] Brender JD, Felkner M, Suarez L, Canfield MA, Henry JP. Maternal pesticide exposure and neural tube defects in Mexican Americans. Annals of epidemiology. 2010;20(1):16-22.

[23] Shaw GM, Nelson V, Olshan AF. Paternal occupational group and risk of offspring with neural tube defects. Paediatric and perinatal epidemiology. 2002;16(4):328-33.

[24] Fear NT, Hey K, Vincent T, Murphy M. Paternal occupation and neural tube defects: a case–control study based on the Oxford Record Linkage Study register. Paediatric and perinatal epidemiology. 2007;21(2):163-8.

[25] Schmidt RJ, Romitti PA, Burns TL, Browne ML, Druschel CM, Olney RS. Maternal caffeine consumption and risk of neural tube defects. Birth Defects Research Part A: Clinical and Molecular Teratology. 2009;85(11):879-89.

[26] Crider KS, Cleves MA, Reefhuis J, Berry RJ, Hobbs CA, Hu DJ. Antibacterial medication use during pregnancy and risk of birth defects: National Birth Defects Prevention Study. Archives of pediatrics & adolescent medicine. 2009;163(11):978-85.

[27] Goetzinger KR, Stamilio DM, Dicke JM, Macones GA, Odibo AO. Evaluating the incidence and likelihood ratios for chromosomal abnormalities in fetuses with common central nervous system malformations. American journal of obstetrics and gynecology. 2008;199(3):285. e1-. e6.

[28] Pilu G, Falco P, Gabrielli S, Perolo A, Sandri F, Bovicelli L. The clinical significance of fetal isolated cerebral borderline ventriculomegaly: report of 31 cases and review of the literature. Ultrasound in Obstetrics and Gynecology: The Official Journal of the International Society of Ultrasound in Obstetrics and Gynecology. 1999;14(5):320-6.

[29] Forzano F, Mansour S, Ierullo A, Homfray T, Thilaganathan B. Posterior fossa malformation in fetuses: a report of 56 further cases and a review of the literature. Prenatal Diagnosis: Published in Affiliation With the International Society for Prenatal Diagnosis. 2007;27(6):495-501.

[30] Ekin A, Gezer C, Taner C, Ozeren M, Ozer O, Koç A, et al. Chromosomal and structural anomalies in fetuses with open neural tube defects.

Journal of Obstetrics and Gynaecology. 2014;34(2):156-9.

[31] Ballarati L, Rossi E, Bonati MT, Gimelli S, Maraschio P, Finelli P, et al. 13q Deletion and central nervous system anomalies: further insights from karyotype–phenotype analyses of 14 patients. Journal of Medical Genetics. 2007;44(1):e60-e.

[32] Purves D. Principles of neural development: Sinauer Associates, Incorporated; 1985.

[33] Recklinghausen Fv. Untersuchungen über die Spina bifida: Walter de Gruyter GmbH & Co KG; 2018.

[34] Padmanabhan R. Experimental induction of cranioschisis aperta and exencephaly after neural tube closure: a rat model. Journal of the neurological sciences. 1984;66(2-3):235-43.

[35] Morgagni G. The Seats and Causes of Diseases, Investigated by Anatomy, 3 vols., trans. Benjamin Alexander (London: A Millar et al, 1769).1(652):48-9.

[36] Kehl S, Schelkle A, Thomas A, Puhl A, Meqdad K, Tuschy B, et al. Single deepest vertical pocket or amniotic fluid index as evaluation test for predicting adverse pregnancy outcome (SAFE trial): a multicenter, open-label, randomized controlled trial. Ultrasound in Obstetrics & Gynecology. 2016;47(6):674-9.

[37] Rossi A, Prefumo F. Additional value of fetal magnetic resonance imaging in the prenatal diagnosis of central nervous system anomalies: a systematic review of the literature. Ultrasound in Obstetrics & Gynecology. 2014;44(4):388-93.

[38] Smith FW, Adam AH, Phillips WP. NMR imaging in pregnancy. The Lancet. 1983;321(8314):61-2.

[39] Kandula T, Fahey M, Chalmers R, Edwards A, Shekleton P, Teoh M, et al. Isolated ventriculomegaly on prenatal ultrasound: what does fetal MRI add? Journal of medical imaging and radiation oncology. 2015;59(2):154-62.

[40] Griffiths P, Porteous M, Mason G, Russell S, Morris J, Fanou E, et al. The use of in utero MRI to supplement ultrasound in the foetus at high risk of developmental brain or spine abnormality. The British Journal of Radiology. 2012;85(1019):e1038-e45.

[41] Garcia-Flores J, Recio M, Uriel M, Canamares M, Cruceyra M, Tamarit I, et al. Fetal magnetic resonance imaging and neurosonography in congenital neurological anomalies: supplementary diagnostic and postnatal prognostic value. The Journal of Maternal-Fetal & Neonatal Medicine. 2013;26(15): 1517-23.

[42] Gonçalves L, Lee W, Mody S, Shetty A, Sangi-Haghpeykar H, Romero R. Diagnostic accuracy of ultrasonography and magnetic resonance imaging for the detection of fetal anomalies: a blinded case–control study. Ultrasound in Obstetrics & Gynecology. 2016;48(2):185-92.

[43] Shapira-Zaltsberg G, Grynspan D, Quintana M, Dominguez P, Reddy D, Davila J, et al. MRI features of the placenta in fetuses with and without CNS abnormalities. Clinical radiology. 2018;73(9):836. e9-. e15.

[44] Hsiao EY, Patterson PH. Placental regulation of maternal-fetal interactions and brain development. Developmental neurobiology. 2012;72(10):1317-26.

[45] Siauve N, Chalouhi GE, Deloison B, Alison M, Clement O, Ville Y, et al. Functional imaging of the human placenta with magnetic resonance. American journal of obstetrics and gynecology. 2015;213(4):S103-S14.

[46] Bonel HM, Stolz B, Diedrichsen L, Frei K, Saar B, Tutschek B, et al. Diffusion-weighted MR imaging of the placenta in fetuses with placental insufficiency. Radiology. 2010;257(3):810-9.

[47] Vedmedovska N, Rezeberga D, Teibe U, Melderis I, Donders GG. Placental pathology in fetal growth restriction. European Journal of Obstetrics & Gynecology and Reproductive Biology. 2011;155(1):36-40.

[48] Pomorski M, Zimmer M, Florjanski J, Michniewicz J, Wiatrowski A, Fuchs T, et al. Comparative analysis of placental vasculature and placental volume in normal and IUGR pregnancies with the use of three-dimensional Power Doppler. Archives of gynecology and obstetrics. 2012;285(2):331-7.

[49] Anencephaly* MTFo. The infant with anencephaly. New England Journal of Medicine. 1990;322(10):669-74.

[50] Thompson DN. Postnatal management and outcome for neural tube defects including spina bifida and encephalocoeles. Prenatal Diagnosis: Published in Affiliation With the International Society for Prenatal Diagnosis. 2009;29(4):412-9.

[51] Caldarelli M, Rea G, Cincu R, Di Rocco C. Chiari type III malformation. Child's Nervous System. 2002;18(5):207-10.

[52] Claridge EA, Bloemen MA, Rook RA, Obeid J, Timmons BW, Takken T, et al. Physical activity and sedentary behaviour in children with spina bifida. Developmental Medicine & Child Neurology. 2019;61(12):1400-7.

Chapter 7

Congenital Diaphragmatic Hernia: A Major Challenge for Neonatologists

Rameshwar Prasad

Abstract

Congenital diaphragmatic hernia (CDH) is a major congenital anomaly of the neonates, characterized by the herniation of abdominal contents into the thoracic cavity during fetal life. This results in significant pulmonary hypertension and hypoxemia after birth, which responds poorly to therapeutic interventions. CDH is associated with high morbidity and mortality. The exact pathogenesis is not well understood, and genetic factors have been proposed. The management starts in utero, with antenatal diagnosis and identification of prenatal predictors for the outcomes, which help in the selection of cases suitable for fetal therapy. The postnatal management is complicated by the need for variable cardio-respiratory support and even extra corporeal membrane oxygenation (ECMO), before corrective surgery is undertaken. Improvement in the understanding of the pathophysiology of the underdeveloped lungs and pulmonary vessels has contributed to substantial progress in the management of CDH, which has translated into improved outcomes and survival. Still, many questions regarding CDH remain unanswered and the management is largely based on weak evidence.

Keywords: congenital diaphragmatic hernia, fetoscopic endoluminal tracheal occlusion, hypoxemia, pulmonary hypertension, diaphragmatic eventration, pulmonary hypoplasia

1. Introduction

CDH is a rare and major congenital anomaly. It is characterized by the partial or complete absence of diaphragm on one or both sides with herniation of abdominal content into the thorax. The dual-hit hypothesis, proposed by Keijzer et al, has suggested that the early insult in lungs' development is bilateral, occurring before and independent of the diaphragmatic defect; and a later second hit to the ipsilateral lung via compression from the herniated abdominal content leads to the characteristic pulmonary hypertension and hypoxemia in the neonate [1].

Advanced prenatal evaluation and a multidisciplinary perinatal management approach have contributed significantly to the improvement in the outcome of CDH. Several prognostic indicators have been forwarded in an attempt to identify candidates with better outcome potential who could benefit maximally from the antenatal and postnatal interventions. CDH is considered to be a medical emergency and the initial intensive management is determined by the severity of the cardiorespiratory failure, which is a consequence of lungs hypoplasia, pulmonary vascular maldevelopment, and the ventricular dysfunction, the three outstanding pathological features of the anomaly. The surgical repair is generally deferred by consensus among the neonatologists and the surgeons. The management is more complicated if CDH is associated with other organ anomalies, which might make the outcome worse. CDH is one of the most challenging morbidities in the neonates. In this article, an evidence based overview of the current status of the disease entity is provided.

2. Epidemiology

The reported prevalence of CDH varies among studies [2–5]. The European Surveillance of Congenital Anomalies (EUROCAT) registry data analysis (1980-2009) has described the prevalence to be 2.3 per 10,000 births for all-inclusive and 1.6 per 10,000 for isolated cases [6]. Balayla *et al*. reported an incidence of 1.93 per 10,000 births in the United States [7]. The variability in the reported prevalence is due to differences in the studied geographical population, data collection methodologies, inclusion and exclusion criteria, case ascertainment and hidden mortality [8]. A decreasing trend in the live births with CDH has been reported, most likely due to an increasing number of termination of pregnancies with antenatal diagnosis of fetuses with CDH [5, 6]. Overall, the survival rate in CDH, although variable among centers, is >70% and has consistently improved over time [9, 10]. A non-significant improvement in the survival of CDH cases which are complicated with major cardiovascular or chromosomal anomalies has been recently reported [10]. The implementation of the standardized management protocols has improved the survival rate from 67% to 88% [11]. Overall, the evaluation of epidemiology and risk factors for CDH is arduous due to the heterogeneity among the studies.

CDH is more common in males [7]. The reports regarding the association of high maternal age with increased risk of CDH are conflicting [5–7, 12]. Pre gestational hypertension [5, 13] and alcohol abuse are other proposed maternal risk factors [7, 13, 14]. The impacts of ethnicity, race, maternal tobacco use and pre gestational diabetes on the risk for CDH are unclear and need further research, although a slightly lower occurrence in blacks has been documented. The identification of modifiable antenatal risk factors help in deciding the direction of prenatal screening and thus the prevention of CDH.

3. Classification of CDH

Congenital diaphragmatic defects (CDD) are classified as posterolateral (Bochdalek, 70-75%) and non-posterolateral according to the location. Non-posterolateral hernias can be retrosternal (Morgagni-Larrey, 23-28%) or central (2-7%), that involves the non-muscular or central tendinous portion of the diaphragm [15]. This anatomical classification has drawbacks as the actual site of lesion may not be easy to discern. Around 85% of the posterolateral CDH are on the left, while 10% are right sided and 5% bilateral. The diaphragmatic eventration is a rare anomaly in which a part or whole of the hemi diaphragm is abnormally elevated into the thoracic cavity, as the normal diaphragmatic musculature is partly or fully replaced by a thin fibro membranous membrane [15], allowing the abdominal viscera to protrude upwards. In order to elucidate the developmental pathways of diaphragmatic defects more accurately, Ackerman et al developed a phenotype worksheet to capture the precise morphological data on CDD by retrospectively analyzing autopsies of 181 cases [16]. They proposed a new classification system based on the location and other specifications. According to the fetal imaging CDH can also be classified as intrapleural and mediastinal; this might help in prenatal counselling [17]. Although inadequately studied, the patho-anatomy of the diaphragmatic defects is relevant for understanding the genetics of CDH, as well as for comparing the outcomes.

Sixty percent of the CDH cases are isolated and not associated with any other major malformations. Pulmonary hypoplasia, gut malrotation, cardiac dextro position and left sided heart hypoplasia are considered to be parts of the sequence of CDH and are not regarded as separate entities. The remaining 40% of complex CDH cases are associated with major congenital anomalies in either syndromic or non-syndromic form.

4. Etiopathogenesis of CDH

4.1 Environmental factors

Environmental factors contributing to CDH are not well investigated. Maternal vitamin A deficiency, nitrofen treatment and retinoid receptor knock-out in animal model are reported to result in CDH in their offsprings, suggesting a role of retinoid signaling pathway in the pathogenesis [18]. Similar effects are seen in WT1 and COUP-TFII mutant mouse models. Beurskens et al found a significantly higher risk of CDH in the infants born to mothers who had low vitamin A intake during pregnancy (Odds ratio, 7.2; 95% confidence interval, 1.5-34.4; $p = 0.01$) [19]. The cord blood in neonates with CDH is reported to have low retinol and retinol-binding protein levels. Thalidomide and quinine have also been implicated as possible causes of CDH [20].

4.2 Genetics of CDH

CDH is genetically heterogeneous although can be monogenic. A genetic cause is found in 30% of the cases, even though most cases are sporadic and about 1-2% familial [21]. The application of the next generation sequencing technologies, e.g. DNA array and whole genome sequencing has contributed to the identification of cases with genetic etiology.

Chromosomal anomalies account for about 10% of the CDH cases. The common chromosomal aneuploidies associated with CDH are trisomy 18, 13 and 21. Morgagni hernia is more frequently found in trisomy 21 than Bochdalek hernia. Pallister Killian syndrome (PKS), characterized by mosaic tetrasomy 12p is a common chromosomal anomaly associated with CDH and occurs sporadically. The karyotype in PKS is often found to be normal due to a tissue specific distribution of chromosome i12p and reduced yield with culture aging [22, 23]. Establishing diagnosis in such cases depends on the tissue examined and genetic test used [24]. Isochromosome 12p is rarely isolated from cord blood lymphocytes, whereas its yield from the skin fibroblast is close to 100% [25]. Also, the chorionic villus sampling may miss mosaicism, while detection rate of amniocentesis is nearly 90% [23, 26]. Chromosome 12p targeted-FISH, array comparative genomic hybridization (aCGH) or other newer genetic technologies should be used to prevent misdiagnosis in the condition [24, 27]. The availability of information regarding genetic etiology equips the clinicians with better understanding of the prognosis for discussion with the parents.

Some of the common monogenic syndromes associated with CDH are Cornelia de Lange, Donnai Barrow and Simpson-Golabi-Behmel, while, others, like Fryn syndrome, Pentalogy of Cantrell and thoracoabdominal syndrome are of unknown etiopathogenesis. Fryn syndrome, a close differential of PKS, is a clinical diagnosis and is associated with CDH in 80% of the cases. Other features of this syndrome are nail hypoplasia, high arched or cleft palate and a characteristic facies that is similar to PKS. Excellent reviews on genetics in CDH are available elsewhere [21, 28, 29].

4.3 Pathology and clinical presentation

The diaphragm, made primarily of muscle, connective tissue and central tendon, develops from the septum transversum, pleuroperitoneal folds, and the somites. The development is complete by 8 weeks of gestation. The defect in CDH is due to an abnormal development of diaphragm during the embryonic phase. Human post mortem reports and animal studies have demonstrated that in CDH, both lungs are hypoplastic, the ipsilateral one being more that the contralateral lung. Characteristically the lungs have decreased DNA and protein contents; diminished airway generations, terminal bronchioles and alveolar volume; thickened alveolar septum, and decreased complexity of the respiratory acinus. There is thickening of the pulmonary arterial medial wall and muscularisation of the smaller pre-acinar arteries.

CDH is mostly diagnosed in utero with ultrasonography (USG), magnetic resonance imaging (MRI), or both. However, some cases may present at birth without prenatal diagnosis. Clinically, the neonate develops respiratory distress, at times severe, at birth or within the first 24 hours of life. If the defect is small there may not be significant respiratory compromise. On physical examination, typically,

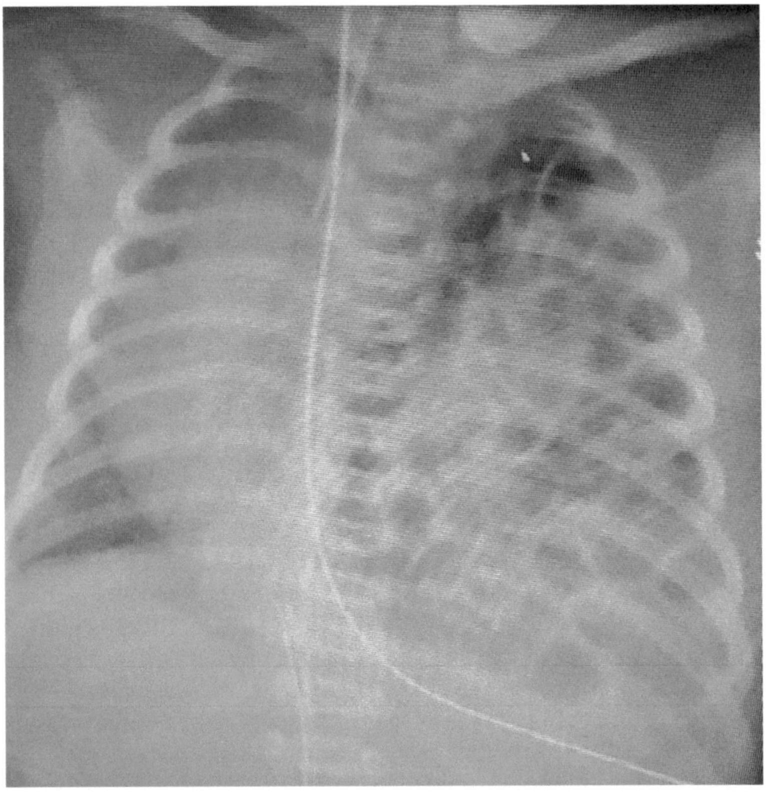

Figure 1.
Chest x-ray showing left sided diaphragmatic hernia in a newborn infant.

breath sounds are decreased on the affected side; abdomen is scaphoid; bowel sounds are heard in the thorax and heart may be displaced towards the contralateral side. Chest radiography is diagnostic (**Figure 1**).

5. Prenatal assessment of CDH

5.1 Prenatal counselling to parents

Prenatal counseling is provided after a comprehensive assessment is made with the information obtained via advanced genetic testing, radio imaging and individualized prognosis based risk stratification. It is performed by a team that is experienced in the pre and postnatal management of CDH. A multidisciplinary approach in counseling the parents by specialists from obstetrics, neonatology, pediatric surgery, genetics and radiology is of paramount value and should be undertaken. It is imperative that an accurate prediction of the outcome is made. The goal is to help the parents in making crucial decision on the options such as, termination of pregnancy (TOP), fetal intervention and expectant management. It also provides guidance for the postnatal management of CDH. Once prenatally diagnosed, mothers should be referred to a tertiary care center that provides expertize in the pre and postnatal management of the neonatal disease.

5.2 Prenatal diagnosis and evaluation of outcomes

Antenatal ultrasound screening may identify >70% of the cases of CDH. The antenatal diagnosis is particularly difficult before 24 weeks of gestation [30, 31] when small, or right-sided hernia may be often missed. A standardized assessment of CDH via prenatal ultrasound has been proposed by the European Reference Network on Rare Inherited and Congenital Anomalies (ERNICA) [32]. The most important information sought from the imaging procedures are the assessment of pulmonary hypoplasia, severity of pulmonary hypertension and the presence of associated major congenital anomalies. Absolute volumetry is superior for confirming the diagnosis of CDH in cases with equivocal sonographic findings and can be done by both 3D ultrasound and MRI.

Several imaging parameters are used for the antenatal risk stratification of CDH (**Table 1**). The lung to head ratio (LHR) of the contralateral side, first described by Metkus et al, is dependent on the gestational age [33–37] and its predictive utility on the postnatal outcome in isolated CDH, or the one associated with liver herniation, is controversial. [35, 38–40]. Still, the observed to expected LHR (o/e LHR) is a widely utilized prediction parameter for counseling the parents and selecting patients for fetal therapy. O/e LHR demonstrates little change with gestational age and provides the ability to predict survival in both left and right sided CDH [41]. Based on o/e LHR, the severity of left sided CDH is classified as follows: < 15%-extreme; < 25%-severe; 25–34.9%, or 35–44.9% with intrathoracic liver herniation-moderate; 35–45% without liver herniation or ≥ 46% -mild [42]. An o/e LHR value of < 25% predicts <25% survival after the first year of life in isolated left-sided CDH, compared to 86.7% in those who have an o/e LHR of 36-45% without liver herniation, or if the value is >45% [43]. In the right-sided CDH, an o/e LHR <45% predicts poor outcome [44]. Ultrasound measurement of o/e LHR is recommended to be performed between the 22nd and 32 weeks of gestational age, although it is reported to be accurate in predicting survival even between the 18 and 38 weeks of gestation in CDH [45]. Considering that o/e LHR varies with fetal

Assessment of Lung size **Indirect methods:** • Contralateral LHR[a] [33] • O/E LHR [41] • Quantitative lung index [46] **Lung volumes** • Observed/expected contralateral fetal lung volume ratio [40, 47] • Observed/expected total fetal lung volume ratio [40, 48] • Fetal lung volume/fetal weight ratio [49] • Absolute TFLV[b] [82] • o/e TFLV [50] • PPLV[c] [51] • Fetal lung volume to fetal body volume [52] • Lung/liver signal intensity ratio [53, 54] • MRI Relative Lung-to-Head Ratio [55]	**Evaluation of Pulmonary circulation** • Modified McGoon index [56] • Prenatal pulmonary hypertension index [56] • Pulsatility index [57, 58] • peak early diastolic reversed flow [57–59] • Fetal Pulmonary arterial diameter [60] • Fractional moving blood volume [61] • Hyperoxygenation tests (pulmonary artery Doppler response to maternal hyperoxia) [62] • Observed/expected diameters of pulmonary arteries [63] • Contralateral vascularization index [40] • Pulmonary artery index [64]
Liver herniation • Binary –up/Down [65] • Liver-to-thoracic area ratio (%) [66] • %LH[d] [67] • Liver-to-thoracic volume ratio (%) [68]	**Stomach position** (indirect marker of liver herniation) Grading: Kitnao [69] , Cordier [70], Basta [71] **Presence of Hernial sac** [72, 73]

[a]*Lung-head ratio.*
[b]*Total fetal lung volume.*
[c]*Percent predictive lung volume.*
[d]*Liver herniation.*

Table 1.
Antenatal imaging parameters to predict outcome.

maturation, Quintero et al proposed Quantitative Lung Index (QLI), which is independent of gestational age [46] This parameter however, , failed to show superior accuracy over other predictive criteria in subsequent studies [74].

In the left sided CDH, liver herniation (LH), present in almost 50% of the cases, is an independent prognostic predictor [65]. The predictive value of LH is lost in the right sided CDH as liver is almost always herniated [75]. LH has been described variably by ultrasonography (USG) as either binary (up or down) or liver-to-thorax ratio, whereas, it is evaluated as liver to thoracic volumes ratio (LiTR) and %LH by the magnetic resonance imaging. The presence of LH suggests a larger defect, a greater severity of the lung hypoplasia and a need for prosthetics to repair the defect [76]. Even though the sensitivity and specificity of the binary parameter are reported to be only 73% and 54% respectively [77], a meta-analysis has concluded that survival is significantly better if the liver is down than up (73.7% vs. 45.4%, $p < 0.005$).

Fetal MRI is a more useful procedure for defining the types of hernia. It predicts the outcome more accurately and allows the detection of other associated congenital anomalies [74, 78, 79]. Both lungs can be assessed by MRI. MRI is more reliable than three-dimensional USG in the characterization of the ipsilateral lung and the estimation of fetal lung volumes [80]. Fetal MRI is recommended in the moderate or severe cases, or if the sonographic evaluation is inconclusive [81]. The procedure can measure total fetal lung volume (TFLV), Observed/expected TFLV (o/e-TFLV), percent predicted lung volume (PPLV), fetal lung volume to fetal body volume ratio, lung to liver signal intensity ratio and liver herniation (%LH) [80]. The TFLV estimated during late gestation can estimate the degree of compromise in the fetal lung growth and a value of <20 ml at 34 weeks of gestation is significantly associated with mortality, as well as the need for ECMO [82]. O/e TFLV < 0.25–0.35% [83, 84] and PPLV <15% have been associated with poor prognosis [85, 86]. A combination of MRI o/e-TFLV and %LH has the best reported accuracy (AUC 0.83, range 0.70−0.91) in predicting mortality and the need for ECMO [87]. The cut off values of o/e-TFLV <35% and that of %LH >20% are associated with increased risks for both death and pulmonary morbidities [88, 89]. Three-dimensional reconstruction of the defects in CDH by MRI is a promising radiological procedure to be utilized in the evaluation and management [90].

The status of fetal pulmonary circulation has been utilized to assist in the prediction of neonatal outcomes in CDH. In fetuses undergoing FETO, the fetal pulmonary vascular reactivity to maternal oxygen supplementation (Δpulsatility index, ΔPI) and the observed/expected lung-to-head ratio (o/e LHR) are independent predictors of pulmonary hypertension and neonatal survival [57, 62]. These parameters, along with peak early diastolic reversed flow (PEDRF) help in identifying the subgroups that may have very poor outcome with fetoscopic endoluminal tracheal occlusion (FETO) [57]. O/e-LHR (>26%, Odds ratio 14.2, survival rate 90%), intrapulmonary artery PI (<1 Z score, Odds ratio 8.4) and PEDRF (<3.5 Z score, Odds ratio 5.7) are associated with survival. O/e LHR of < 26% predicts a survival rate of only 45%. If fetal o/e LHR is < 26%, Doppler parameters can identify cases who would have chances of survival between 0% to 66-71%. Other fetal MRI parameters, such as, McGoon index and pulmonary artery index, have also been used to predict the mortality [64].

The prenatal evaluation tests, however, poorly predict the severity, duration, and response to therapy of the postnatally appearing persistent pulmonary hypertension (PPH) in CDH. Spaggiari et al documented the association between the presence and degree of pulmonary hypoplasia and post natal PPH in CDH [91]. Modified McGoon index (MMI) and prenatal pulmonary hypertension index (PPHI) are reported to accurately predict the severity of PPH [56]. Done et al have demonstrated that the fetal pulmonary reactivity to maternal O_2 administration (ΔPI) and liver-to-thorax ratio (LiTR) are the best predictors of PPH that would last for ≥28 postnatal days [92]. Notwithstanding, a recent meta-analysis failed to establish the predictive utility of any of these prenatal parameters for PPH in CDH cases [37].

The presence of a hernia sac predicts better survival [72, 73]. The location of fetal stomach, such as intraabdominal, anterior left chest, posterior-mid left chest or retro cardiac, has been shown to strongly predict neonatal outcomes in isolated left CDH [69–71]. An abnormal stomach position is associated with mortality (OR 4.8, 95% CI 2.1-10.9), ECMO requirement (OR 5.6, 95% CI 1.9-16.7), non-primary diaphragmatic repair (OR 2.7, 95% CI 1.4-5.5), and extended mechanical ventilation (OR 5.9, 95% CI 2.3-15.6), whereas, the presence of an intra-abdominal stomach predicts survival without significant respiratory morbidity or without the need for ECMO. Prenatal intra fetal fluid effusions, commonly observed in the antenatal USG in fetuses presenting with left and right-sided CDH, with the occurrence rates of 5% and 29% respectively, do not have any association with poor outcomes [93]. In an attempt to improve the predictive value of individual parameters, 10 prenatal tests were combined into a single congenital diaphragmatic hernia composite prognostic index (CDH-CPI) by Le et al, which is reported to be strongly associated with both, the survival and the need of ECMO in infants with CDH [94]. Estimation and Risk stratification at the time of diagnosis of CDH via scans at 18-20 weeks of gestation that could provide an accurate assessment of the postnatal outcomes, would be the ideal tool to determine whether or not a timely termination of pregnancy should be undertaken.

5.3 Postnatal evaluation of outcomes

The neonatal survival in CDH, to a large extent depends on the postnatal management provided to the infant after birth. CDH may not be always diagnosed prenatally and may be detected unexpectedly at birth. Furthermore, in a prematurely born fetus, the antenatal predictions of the prognosis become non-applicable as a significant new risk factor for the outcome and survival is added. This necessitates a new evaluation at birth [94]. Most postnatal prediction tools have been derived from and validated by the large registries. The intraoperative defect size staging, and the presence of associated anomalies, together are probably the most reliable predictor of the severity of the morbidity as well as of the survival [95, 96]. Congenital Diaphragmatic Hernia Study Group (CDHSG) developed a staging system based on the size of the diaphragmatic defect, as detected intraoperatively during the diaphragmatic repair, and the presence or absence of major cardiac anomalies [97]. In their study, the stage I patients with isolated small defects had 99% survival rate, compared to only 39% for stage V patients who had large diaphragmatic defects associated with major cardiac anomalies; and 0% in the group that was not reparable [98]. A major limitation of this staging system is the fact that being an intraoperative classification method it cannot be applied to infants before operation or to those who do not survive for the surgery. Other postnatal prediction tools for the outcomes are the Wilford Hall/Santa Rosa clinical prediction formula (WHSRPF) [99], CDH study group predictive survival [100], arterial blood gas parameters [101–103], the Score for Neonatal Acute Physiology, Version II (SNAP-II) [104], the Score for Neonatal Acute Physiology Perinatal Extension (SNAPPE-II) [105] and Brindle score [106, 107].

Cardiovascular defects are the most common anomalies associated with CDH [6, 108]. A meta-analysis involving 28,974 CDH cases reported the occurrence rate of cardiovascular defects to be 15%, out of which 42% were of critical nature [109]. All infants with CDH must undergo a high-quality diagnostic echocardiogram during pregnancy, and a postnatal imaging should be performed at birth regardless of their antenatal imaging findings. About 61% of the cases of the commonest types of CDH present with other malformation, which include chromosomal defects, non-chromosomal syndromes, malformation sequences, malformation complexes and non-syndromic multiple congenital anomalies [108]. Malformations of the cardiovascular system (27.5%), urogenital system (17.7%), musculoskeletal system (15.7%), and central nervous system (9.8%) are the commonest congenital anomalies noted in CDH. A recognizable malformation syndrome may be identified in 57.1% of the cases of CDH.

6. Antenatal interventions to improve outcome

Antenatal steroids (ANS, betamethasone and dexamethasone) are reported to improve the morphological, physiological and biochemical indicators of pulmonary immaturity, and effectuate the vascular remodeling in animal models of CDH [110, 111]. ANS variably affects the surfactant production in CDH [112, 113]. Researchers have demonstrated contrasting effects of ANS on the tracheal occlusion procedure (TO). The fetal lung growth acceleration by TO is noted to be countered by steroids [114]. On the other hand, TO adversely affects the type II pneumocytes, whereas,

ANS increases type II pneumocytes cell density and SPmRNA expression [115], thus partially correcting the surfactant deficiency caused by the procedure [116]. Combined TO and betamethasone treatment might have additive effects on reducing the muscularization of pulmonary arteries in hypoplastic lungs of CDH [117]. However, although the expression of vascular endothelial growth factor receptors is found to be restored by both TO and dexamethasone individually, the results are not noted to be additive. The antenatal betamethasone therapy was studied in a randomized controlled human trial by administering two doses of 12.5 mg, given 24 hours apart, followed by 12.5 mg given weekly for 2 weeks to pregnant women carrying CDH fetuses of 34 weeks of gestation. An interim analysis of the 32 enrolled cases showed no differences in the perinatal mortality, the duration of mechanical ventilation or the length of hospitalization between the treated and placebo groups. The study ended prematurely, as it was calculated that >1700 prenatally diagnosed CDH pregnancies would be needed to be enrolled in order to show a 10% improvement in the survival. CDH registry also failed to show any beneficial effects of the late gestation steroids therapy in CDH fetuses [118]. In the light of beneficial effects of ANS in reducing the respiratory morbidities in late preterm babies [119], Canadian CDH guideline recommends a single dose of antenatal steroids until the gestational age of 36 weeks is achieved [81]. A large multicenter trial is needed to be conducted in order to resolve the issues related to the role of ANS in CDH.

Persistent pulmonary hypertension (PPH) is one of the main causes of mortality and morbidity in CDH. Prenatal sildenafil administration in the expectant mothers has been shown to prevent fetal and neonatal vascular changes that lead to pulmonary hypertension in animal models [120]. The beneficial effects of antenatal sildenafil on neonatal pulmonary hemodynamics in lambs with diaphragmatic hernia (DH) were demonstrated in another study [121]. Conversely, sildenafil is found to decrease the lung-vessel density and vascular branching in normal lungs without CDH in animal models, raising concerns about its safety in human fetuses [120, 122]. Antenatal sildenafil has been investigated for fetal benefits in several randomized controlled trials for the prevention of preeclampsia, fetal growth restriction (FGR) and idiopathic oligohydramnios [123]. The multicenter STRIDER trial (Sildenafil Therapy in Dismal Prognosis Early Onset Fetal Growth Restriction) didn't show any benefit of antenatal sildenafil in FGR, although no adverse effects were attributed to the drug [124]. The recent meta-analyses suggest the use of sildenafil during pregnancy to be safe and to be associated with improved fetal weight at birth [123, 125]. However, the safety data for the use of antenatal sildenafil on human fetuses with CDH are scant. In view of the lack of safety data, a phase I/IIb placental transfer and safety study in human fetuses afflicted with CDH is currently underway (Antenatal Sildenafil Administration to Prevent Pulmonary Hypertension in Congenital Diaphragmatic Hernia, SToP-PH) [126].

The drug, retinoic acid crosses placenta and improves lung maturation in the nitrofen rat model of CDH [127]. However, more information is needed about its teratogenic effects before human trials can be undertaken. Other potential drugs, such as, Imatinib, [128], Vitamin E [129], bombesin [130] and stem cell based therapies [130, 131] are still limited to preclinical trials.

7. Postnatal management

The decision about the location of delivery in centers with high versus low volumes of CHD cases is controversial. Grushka et al. observed that the survival was higher in infants who were born in centers that had > 6 cases per year [132]. However, the evidence in the study was weak. There is no evidence supporting any positive relationship between the numbers of available specialized surgeons and the morbidity related outcomes [133]. It has been documented that the out born neonates have higher odds of mortality (Odd's ratio 2.8, P=0.04) than those who are inborn [134]. The transfer of pregnant mothers with fetus in-utero is strongly recommended and must be done as often as possible. Current data support the decision that the prenatal care and delivery should be conducted at an optimally equipped hospital, which would be capable of providing the best care for best outcomes in newborns suffering from CDH [135]. However, the VICI-trial (High Frequency Oscillation Versus Conventional Mechanical Ventilation in Newborns with Congenital Diaphragmatic Hernia) observed similar survival rate between patients born at ECMO or at non-ECMO centers [136].

The timing of delivery is controversial and decided by the caretaking team as per the case. The gestational age at birth is a strong predictor of outcome [137]. Scheduled delivery at ≥ 39 weeks, or otherwise as per the maternal indications is considered to be the best approach in the wake of current knowledge. An analysis of 928 CDH neonates born vaginally after spontaneous onset of labor, showed a reduction in the mortality if the delivery occurred at 40 weeks of gestation (16.7%) compared to 37 weeks (25%) [138]. The Canadian Pediatric Surgery Network (CAPSNet) studied prenatally diagnosed cases of CDH under three gestational age groups (<37 weeks, 37-38 weeks, >39 weeks) and observed no differences in the mortality, ECMO requirement, duration of mechanical ventilation, the length of stay, or dependence on supplemental oxygen at discharge [139]. The outcome of CDH is worse in the prematurely born infants. It is demonstrated that the FETO procedure puts preterm neonates at a higher risk for mortality and morbidity [140]. Poor neurodevelopmental outcome has been reported in CDH survivors who are born prematurely [141]. The CAPSNet data [139] and CDH Study Group (CDHSG) reports [137] do not support any preferred delivery mode, i.e. vaginal or Cesarean section.

The resuscitation of the antenatally diagnosed cases of CDH should be performed according to the neonatal resuscitation guidelines from the American Heart Association [142] and the American Board of Pediatrics. These are endorsed by both the CDH EURO Consortium [143] and Canadian CDH collaboration [81]. The Neonatal Resuscitation Program (NRP) recommends immediate endotracheal intubation and avoidance of bag-mask ventilation in neonates with CDH. An orogastric tube should be inserted with continuous or intermittent suctioning to prevent bowel distension. A T-piece resuscitator should be used and peak inspiratory pressure (PIP) > 25 cm H2O should be avoided in order to prevent lung injury. The European guidelines also suggest a trial of spontaneous breathing in cases where good lung volume was predicted in the prenatal assessment [143]. In a series of five CDH patients with good lung development (LHR >2.5 or o/e LHR >50%, liver down), it was found feasible to avoid the routine intubation after birth [144].

Appropriate timing of the umbilical cord clamping (UCC) in CDH has not been studied well. A feasibility trial (delayed vs immediate UCC) showed that intubation and ventilation before UCC in CDH are both safe and feasible and the neonates with delayed UCC have significantly higher hemoglobin and mean blood pressure at 1 hour of age than controls [145]. However, overall, the recommendations regarding the delivery room management of CDH cases are not evidence based and more research is needed on this issue. As for the recommendations on the preductal arterial saturations targets in CDH that would ensure adequate tissue perfusion and oxygenation, the CDH EURO consortium recommends preductal SpO 2 in the delivery room to be between 80–95%. If supplemental oxygen is used at resuscitation, it should be rapidly titrated according to the specific preductal oxygen saturation target.

The surfactant status of the lungs in human fetuses with CDH is controversial and needs more information. Studies in the animal models have demonstrated surfactant to be deficient in the lungs with CHD, which might further complicate the pathophysiology of the disease process. In one study, the surfactant amount or its maturation were not found to be deficient in CDH lungs [146], while in a separate study that measured surfactant components in the tracheal aspirates of CDH neonates, both the synthesis rate and amount of SP-B were detected to be lower [147]. A retrospective analysis of 522 term infants who suffered from CDH and were enrolled in the CDHSG registry, showed an increase in the mortality, the need for ECMO, and the occurrence of chronic lung disease in those 122 infants who were treated with exogenous surfactant [148]. Furthermore, retrospective analyses published by the CDH Study Groups found no significant advantage of surfactant when given in the preterm infants with <37 weeks of gestation, and in infants on ECMO who were of > 35 weeks of gestational age, raising questions over its routine use [149, 150]. A meta-analysis also did not show any benefit of postnatal surfactant therapy in CDH cases [151]. Regardless, surfactant is used in the preterm infants with CDH almost invariably, and the trends in this matter have been stable [152]. It is reasonable to think that some specific subsets of neonates with CDH could benefit from the postnatal surfactant administration. These issues need to be evaluated via prospective trials in the future.

The strategies of permissive hypercapnia and 'gentle ventilation' have been demonstrated to improve the outcomes of neonates with CDH [151, 153, 154]. Ventilator induced lung injury caused by high ventilator settings could further damage the hypoplastic lungs and worsen the pulmonary hypertension. A retrospective analysis by Wung, et al has reported improved survival in CDH patients who were managed with minimal sedation, low peak pressures, permissive hypercapnia, avoidance of paralytics and delayed surgery [155]. A prospective study of 120 CDH neonates treated with permissive hypercapnia, permissive hypoxemia, spontaneous respiration and elective surgery found the survival to discharge rate to be at 84.4% [156]. In these cases, the oxygenation targets were set at lower levels as long as other blood gas parameters remained in acceptable range and the markers for adequate tissue perfusion were maintained within the normal values [157]. In a retrospective analysis of 70 high risk CDH patients, the survival significantly improved from 47% to 90% with the routine use of permissive hypercapnia and gentle ventilation [158].

The controversy regarding the optimal preductal and postductal saturation targets is still unresolved and more prospective trials are needed to address the issues. In the first 2 hours after birth, a preductal SpO2 of 70% is adjudged to be acceptable, if it is improving without ventilator changes, is associated with a pH >7.2 suggesting satisfactory organ perfusion, and the ventilation is adequate with PaCO2 below 65 mm Hg. The pulmonary vessels responds to hypoxia by vasoconstriction but hyperoxia has not been shown to result in vasodilation; instead free oxygen radical injury may ensue in such conditions. For these reasons, the SpO2 is preferred to be kept between 90-95% in pulmonary hypertension [81].

The VICI trial provided insight into the optimum initial ventilation management of CDH patients [136]. In this study 171 antenatally diagnosed CDH patients were randomized to be initiated at either conventional ventilation (CMV) or high frequency ventilation (HFO) within 2 hours after birth. Even though, the primary outcome (death/bronchopulmonary dysplasia at day of life 28) was similar in both groups, the conventional mechanical ventilation group had significantly shorter duration of ventilation; as well as, less need for inhaled iNO, sildenafil or ECMO; shorter duration of vasoactive medication and less occurrence of treatment failure. In this crossover study, the neonates were switched over to the other ventilation mode or put on ECMO if the signs of failure of initial ventilation mode occurred and persisted for 3 hours or more. Although the trial was terminated prematurely and only a total of 171 patients were enrolled against the desired sample size of 200 per group, the result of this trial suggested an advantage of initiating CMV over HFOV in regards to several important secondary outcomes.

It is recommended that, along with minimal stimulation, a judicious use of routine sedation and analgesia should be exercised during mechanical ventilation in CDH. Morphine sulfate and fentanyl are the preferred drugs for such purposes. Deep sedation and muscle relaxation impair respiratory functions in newborns with CDH and should be avoided [159]. The validated tools for neonatal pain assessment e.g. Neonatal Pain, Agitation and Sedation Scale (N-PASS), COMFORTneo and COMFORT behavioral scales evaluate both analgesia and sedation [160–162] and help in guiding the therapy as well as in monitoring the effects of medications.

Timing of delivery	Scheduled delivery at $\geq$ 39 weeks or at obstetric indication
Preferred delivery mode	No evidence of any preferred mode: vaginal or caesarean section.
Resuscitation	• Immediate endotracheal intubation and avoidance of bag-mask ventilation • Use T-piece resuscitator and peak inspiratory pressure $\leq$ 25 cm H_2O
Timing of the UCC[a]	Insufficient data
Surfactant	Avoid routine use
Preductal SpO_2	• In deliverey room SpO_2 80-85% • First 2 hours after birth: SpO_2 70% acceptable if pH>7.2, $PaCO_2$ <65 mm Hg.
Optimum initial ventilation	Conventional ventilation, rescue HFOV[b]
Sedation and Analgesia	• Use Judiciously • Use pain and sedation assessment scales
Fluid management	Fluid boluses restricted to 10-20 ml/kg.
Inotropes	• Improve blood pressure without improving microcirculation
Prenatal predictors of PPH[c]	Inconsistent results
Increasing SVR[d]	Not recommended to reverse right to left shunt by increasing SVR
Severe PPH + right to left atrial and ductal shunting	• Maintain the patency of ductus arteriosus with prostaglandin E1 • May respond to iNO[e]
LV dysfunction + left-to-right atrial shunting + right to left ductal shunting	• Impaired LV filling results in elevated left atrial pressure and right to left atrial shunting • Consider pulmonary venous hypertension • Poor response to iNO and potential to develop pulmonary hemorrhage • Milrinone preferred
Resource poor settings/iNO unavailable	Sildenafil – oral / Intravenous

[a] *Umbilical cord clamping.*
[b] *High frequency oscillatory ventilation.*
[c] *Persistent pulmonary hypertension.*
[d] *Systemic vascular resistance.*
[e] *Inhaled nitric oxid.*

Table 2.
Key points for postnatal management of congenital diaphragmatic hernia.

In order to meet the goal of ensuring adequate tissue perfusion the following clinical and biochemical markers should be maintained and monitored while managing a case of CDH: normal heart rates and blood pressure for the gestational age, urine output ≥ 1.0 ml/kg/h and serum lactate concentration <3 mmol/l. The monitoring of intra-arterial blood pressure should be done via an indwelling umbilical artery catheter. If a peripheral arterial line is used, it should preferably be inserted in the right radial artery in order to allow the measurement of pre-ductal PaO2. An echocardiography should be done immediately after or within the first 24 hours of birth and repeated serially as indicated. The left ventricular hypoplasia and cardiac dysfunction are known to be important independent determinants of mortality in CDH and need particular attention [163]. The echocardiography helps in evaluating the myocardial dysfunction, pulmonary hypertension and right to left shunt, in addition to identifying associated structural cardiac anomalies. These information guide medical therapy. Echocardiography helps in differentiating the hypotension due to myocardial dysfunction from that of volume depletion and thus determines the appropriate management in the poor perfusion states. It also evaluates the right ventricular dysfunction and/or right ventricular overload. In CDH left ventricular dysfunction is common and should be monitored for its degree, and be treated appropriately and promptly. Other cardiac complications, such as reduced ventricular compliance and increased vascular resistance may be present and if so contribute to the reduced cardiac output. Poor left ventricles compliance in CDH generates a risk for pulmonary edema with excessive fluid administration [164]. Therefore, judicious fluid resuscitation with fluid boluses restricted to 10-20 ml/kg is advisable. If there is no response after 2 boluses, inotropic and/or vasopressor agents should be strongly considered and started. In cases who are refractory to inotropes, hydrocortisone can be added although its role in CDH is not well established and the subject requires further information. A cohort study involving 28 newborn infants has demonstrated that dopamine, epinephrine and norepinephrine improve blood pressure without improving microcirculation in CDH [165]. There is a dearth of randomized controlled trials comparing inotropes in CDH. The drug milrinone reduces afterload and improves right and left ventricular dysfunction and may be used in selected cases [166]. It is imperative to maintain continuous hemodynamic monitoring of the central venous and arterial pressures, as well as of arterial and mixed venous saturation via appropriately placed catheters. In addition, timely assessment of cardiac functions via serial echocardiographs is an essential and vital components of the management of CDH (**Table 2**).

8. Pulmonary hypertension

Pulmonary hypertension in CDH is persistent, recalcitrant and difficult to treat. The management is compromised by the scarcity of good quality research and evidence. The presence of PPH is the most significant predictor and cause of morbidity and mortality, as well as of the need for ECMO [83]. Pulmonary hypertension in CDH is characterized by the hypoplasia of pulmonary vascular bed, remodeling of pulmonary vasculature, and an altered vasoreactivity of the pulmonary vessels. The pulmonary vasculature remodeling is characterized by the medial and adventitial thickening of the midsized and large vessels' walls and by the neomuscularisation of small capillaries. The vascular remodeling is evidenced to begin at as early as the 19th weeks of gestation in human lungs of CDH patients, and is associated with an earlier maturation of the pulmonary vasculature [167]. The pulmonary hypertension in CDH has both fixed and variable components. Reduced pulmonary vascular cross sectional area and the vascular remodeling contribute to the fixed component of PPH. The reactive component of PPH is balanced by the circulating vasoconstrictor and vasodilator agents and defines the response to vasodilator treatment. Increased expression of the vasoconstrictive factors (endothelin A and B receptors, endothelin converting enzyme-1) and a decreased expression of the vasodilators (prostaglandin-I_2 receptor) have been reported in CDH [168]. The reports about the concentration of endothelial nitric oxide (eNOS) in CDH patients are conflicting [169–172]. An extensive review of the factors and pathways involved in the vascular remodeling in CDH has been published [173].

The utility and selection of pulmonary vasodilator agents for the treatment of PPH often change, as the underlying pathophysiology evolves during the days and weeks following the birth. The management strategies need to be adjusted as the pulmonary vascular abnormalities progress from acute to subacute and finally to the chronic stage. A poor response to the vasodilator therapies during acute phase is attributed to the left ventricular (LV) hypoplasia and dysfunction, in association with pulmonary venous hypertension [174]. In some CDH patients, there may be a "honeymoon" period displaying good oxygenation during the 1st few hours of life. Ratio of the pulmonary arterial to systemic arterial pressure (P/S ratio) is used to evaluate the degree of PPH. A P/S ratio <2/3 of systemic systolic pressures is considered as normal or mild; whereas, that of ≥2/3 of systemic pressure denotes moderate; and if ≥ systemic pressure, severe PPH [175]. Important parameters to check in the neonatal echocardiography are the direction and velocity of flow at ductal and foramen ovale shunts, flattening or left deviation of the interventricular septum, and tricuspid regurgitation jet velocity. Functional evaluation of the right and left ventricles is the key that guides the appropriate pulmonary vasodilator therapy. Lung recruitment with adequate positive end-expiratory (PEEP) or mean airway pressure and/or with surfactant to achieve an expansion up to 8- to 9-ribs during inspiration is the goal. Overriding right-to-left shunt by increasing systemic vascular resistance is not recommended. Severe PPH with right to left shunt at the atrial level necessitates the emptying of right ventricle, which is accomplished by maintaining the patency and reopening of ductus arteriosus with prostaglandin E1 if needed [176, 177]. Such cases may respond favorably to the inhaled nitric oxide therapy (iNO). LV dysfunction is associated with left-to-right atrial shunting and suggests the presence of pulmonary venous hypertension, which is associated with poor response to pulmonary vasodilator therapy [178]. Milrinone is recommended in patients with LV dysfunction due to its lusitropic, inotropic and pulmonary vasodilator properties [174]. A retrospective study reported improved cardiac function and reduced right ventricular pressure with milrinone in PPH associated with CDH [179]. As there is a potential risk of systemic hypotension, the avoidance of loading dose or priming the system with normal saline bolus were suggested by the researchers in some case series studies [180, 181]. Prospective RCTs of milrinone in CDH complicated by PPH are lacking and are warranted [182].

The Neonatal Inhaled Nitric Oxide Study (NINOS), which included 53 infants with CDH showed no difference in the combined primary outcome of death or ECMO utilization between the iNO-treated patients and controls [183]. In fact, ECMO utilization was higher in the iNO-treated group (80% vs 54% control, $p = 0.043$), indicating an adverse effect of iNO if used early in the course of PPH associated with CDH [183]. The ventilation approach, choice of the ventilator and the OI at enrollment in this study were different from the current practices. Their result further supported the important role of left ventricular dysfunction in PPH in CDH cases [178]. The use of iNO was associated with significantly higher mortality in CDHSG registry cases as well [184]. However, despite a lack of evidence, inhaled nitric oxide is commonly used in the management of PPH in CDH [179, 184].

In the nonresponders to iNO, or in the resource poor settings where iNO is often unavailable, sildenafil is a feasible option. Trials to evaluate the effect of sildenafil in newborns with persistent pulmonary hypertension of the newborn (PPHN) have been conducted. A meta analysis showed that sildenafil reduces mortality and improves oxygenation in PPHN [185]. However, there are limited data on the use of sildenafil in CDH patients. In retrospective studies, both oral and IV formulations of sildenafil were shown to improve oxygenation in CDH patients [186, 187], although IV sildenafil resulted in an increased need for inotropic support [188]. The Congenital Diaphragmatic hernia Nitric Oxide versus Sildenafil (CoDiNOS) trial is the first prospective RCT that has been recently proposed [189]. Other drugs under consideration are epoprostenol, treprostinil, inhaled alprostadil and bosentan. At this time, there is insufficiant evidence in regards to their efficacy or safety for use in PPHN [190–192]. It is reasonable to comment that an accurate prediction of the reversibility of PPHN is not possible with the current knowledge.

9. Extracorporeal membrane oxygenation

Congenital diaphragmatic hernia is the most common non-cardiac indication for extracorporeal membrane oxygenation in the neonates, even though the benefits of the intervention in terms of mortality in CDH are questionable. A retrospective study of the extracorporeal life support organization (ECLS) registry data revealed the overall survival of 48.1% in term neonates representing a 13.8% reduction in survival of neonates with CDH who were treated with ECMO [193]. This result probably represents the fact that more serious CDH cases were subjected to ECMO [194]. CDHSG registry data, on the other hand, have documented an improved survival in the most severely affected CDH patients with ECMO [195]. The recently published VICI trial [136] failed to demonstrate any difference in CDH outcomes between ECLS and non-ECLS centers, further questioning the role of ECMO in CDH. The indication and utility of ECMO in CDH is still evolving, as is the timing of surgery. It should be discussed during prenatal counselling and may be considered as a therapeutic option in those centers that offer it.

10. Surgery

The reduction of herniated abdominal contents does not improve PPH in CDH and thus the outcome remains largely unaffected. A systematic review on the subject has not favored either early or late surgery [154]. The recommended physiologic criteria for preparedness for surgery are not supported by evidence [81, 143]. However, It is possible that early repair on ECMO may improve outcome [154]. In stable patients, CDH repair is usually done between 48–72 h after birth. Recurrence rate is significantly lower in open repair [154]. Polytetrafluoroethylene (PTFE) is the most durable patch repair material. Even though the survival in patients requiring patch repair has significantly improved, it is still lower than those who do not require a patch (76.9% vs 96.5%) [10].

11. Follow-up and outcomes

A multidisciplinary approach is needed for the comprehensive long term follow up of the neonates with CDH. A myriad of significant morbidities may affect these neonates, which includes respiratory disorders (chronic lung disease, pulmonary hypertension, obstructive pulmonary disease, reduced exercise capacity, recurrent pulmonary infection), gastroesophageal reflux, nutritional derangements, neurodevelopmental delays, hernia recurrence, hearing deficits and orthopedic deformities [196, 197]. The Health-related Quality of Life (HRQoL) has been reported to improve as the survivors grow older, while it may be variably compromised during the childhood [198–201]. A single-center prospective study evaluated a cohort of 32 CDH survivors at the mean age of 8 ± 4 years and recorded that about 62% of them needed medical equipment, 18% home health services and 28% special education [199]. The HRQoL in the survivors did not differ from that of healthy children, although it was diminished among those who required special education. The study concluded that even though many CDH survivors continue to require home medical equipment and home health services at school age, most have normal parent-reported HRQoL, and that the need for special education and the higher family impact of neonatal CDH lead to decreased HRQoL.

The guidelines for the follow-up of CDH survivors have been outlined by the American Academy of Pediatrics and are available [196]. A review of current follow up practices in CDH EURO consortium centers revealed that even though 15 out of 19 centers had structured and standardized follow up program for the CDH patients, the annual follow up until 16 years of age was not done in any of the participating centers [197]. The study group proposed the implementation of standardized follow-up of CDH patients for extended evaluation of the survivors for their long term outcomes [197].

12. Conclusion

Despite advances in the ante- and postnatal management, CDH is still a major medical and surgical challenge. Major determinants of the outcome in isolated CDH cases are the severity of pulmonary hypertension and concomitant cardiac dysfunction. Postnatal management targeted towards the correction of the underlying right and left ventricular dysfunction may lead the way to improved outcomes in the neonates with CDH. It is anticipated that the translational research and stem cell therapy might revolutionize the management of CDH in the future.

Author details

Rameshwar Prasad
Department of Neonatology, Rajendra Institute of Medical Sciences, Ranchi, Jharkhand, India

*Address all correspondence to: drrameshwarprasad@hotmail.com

References

[1] Keijzer R, Liu J, Deimling J, Tibboel D, Post M. Dual-hit hypothesis explains pulmonary hypoplasia in the nitrofen model of congenital diaphragmatic hernia. Am J Pathol 2000;156:1299-1306. https://doi.org/10.1016/S0002-9440(10)65000-6.

[2] Dott MM, Wong L-YC, Rasmussen SA. Population-based study of congenital diaphragmatic hernia: Risk factors and survival in metropolitan Atlanta, 1968-1999. Birth Defects Res Part A Clin Mol Teratol 2003;67:261-267. https://doi.org/10.1002/bdra.10039.

[3] Wright JCE, Budd JLS, Field DJ, Draper ES. Epidemiology and outcome of congenital diaphragmatic hernia: a 9-year experience. Paediatr Perinat Epidemiol 2011;25:144-149. https://doi.org/10.1111/j.1365-3016.2010.01172.x.

[4] Levison J, Halliday R, Holland AJA, Walker K, Williams G, Shi E, et al. A population-based study of congenital diaphragmatic hernia outcome in New South Wales and the Australian Capital Territory, Australia, 1992-2001. J Pediatr Surg 2006;41:1049-1053. https://doi.org/10.1016/j.jpedsurg.2006.01.073.

[5] Burgos CM, Ehrén H, Conner P, Frenckner B. Maternal Risk Factors and Perinatal Characteristics in Congenital Diaphragmatic Hernia: A Nationwide Population-Based Study. Fetal Diagn Ther 2019;46:385-391. https://doi.org/10.1159/000497619.

[6] McGivern MR, Best KE, Rankin J, Wellesley D, Greenlees R, Addor MC, et al. Epidemiology of congenital diaphragmatic hernia in Europe: A register-based study. Arch Dis Child Fetal Neonatal Ed 2015;100:F137–F144. https://doi.org/10.1136/archdischild-2014-306174.

[7] Balayla J, Abenhaim HA. Incidence, predictors and outcomes of congenital diaphragmatic hernia: a population-based study of 32 million births in the United States. J Matern Neonatal Med 2014;27:1438-1444. https://doi.org/10.3109/14767058.2013.858691.

[8] Harrison MR, Bjordal RI, Langmark F, Knutrud O. Congenital diaphragmatic hernia: the hidden mortality. J Pediatr Surg 1978;13:227-230. https://doi.org/10.1016/s0022-3468(78)80391-1.

[9] Snoek KG, Greenough A, van Rosmalen J, Capolupo I, Schaible T, Ali K, et al. Congenital Diaphragmatic Hernia: 10-Year Evaluation of Survival, Extracorporeal Membrane Oxygenation, and Foetoscopic Endotracheal Occlusion in Four High-Volume Centres. Neonatology 2018;113:63-68. https://doi.org/10.1159/000480451.

[10] Morini F, Lally KP, Lally PA, Crisafulli RM, Capolupo I, Bagolan P. Treatment Strategies for Congenital Diaphragmatic Hernia: Change Sometimes Comes Bearing Gifts. Front Pediatr 2017;5:195. https://doi.org/10.3389/fped.2017.00195.

[11] Antonoff MB, Hustead VA, Groth SS, Schmeling DJ. Protocolized management of infants with congenital diaphragmatic hernia: Effect on survival. J Pediatr Surg 2011;46:39-46. https://doi.org/10.1016/j.jpedsurg.2010.09.060.

[12] Yang W, Carmichael SL, Harris JA, Shaw GM. Epidemiologic characteristics of congenital diaphragmatic hernia among 2.5 million california births, 1989-1997. Birth Defects Res Part A Clin Mol Teratol 2006;76:170-174. https://doi.org/10.1002/bdra.20230.

[13] Mcateer JP, Hecht A, De Roos AJ, Goldin AB. Maternal medical and behavioral risk factors for congenital diaphragmatic hernia. J Pediatr Surg

2014;49:34-38. https://doi.org/10.1016/j.jpedsurg.2013.09.025.

[14] Felix JF, van Dooren MF, Klaassens M, Hop WCJ, Torfs CP, Tibboel D. Environmental factors in the etiology of esophageal atresia and congenital diaphragmatic hernia: Results of a case-control study. Birth Defects Res Part A Clin Mol Teratol 2008;82:98-105. https://doi.org/10.1002/bdra.20423.

[15] Pober BR. Overview of epidemiology, genetics, birth defects, and chromosome abnormalities associated with CDH. Am. J. Med. Genet. Part C Semin. Med. Genet., vol. 145, NIH Public Access; 2007, p. 158-71. https://doi.org/10.1002/ajmg.c.30126.

[16] Ackerman KG, Vargas SO, Wilson JA, Jennings RW, Kozakewich HPW, Pober BR. Congenital Diaphragmatic Defects: Proposal for a New Classification Based on Observations in 234 Patients. Pediatr Dev Pathol 2012;15:265-274. https://doi.org/10.2350/11-05-1041-OA.1.

[17] Mehollin-Ray AR, Cassady CI, Cass DL, Olutoye OO. Fetal MR imaging of congenital diaphragmatic hernia. Radiographics 2012;32:1067-1084. https://doi.org/10.1148/rg.324115155.

[18] Greer JJ, Babiuk RP, Thebaud B. Etiology of congenital diaphragmatic hernia: The retinoid hypothesis. Pediatr Res 2003;53:726-730. https://doi.org/10.1203/01.PDR.0000062660.12769.E6.

[19] Beurskens LWJE, Schrijver LH, Tibboel D, Wildhagen MF, Knapen MFCM, Lindemans J, et al. Dietary vitamin A intake below the recommended daily intake during pregnancy and the risk of congenital diaphragmatic hernia in the offspring. Birth Defects Res Part A - Clin Mol Teratol 2013;97:60-66. https://doi.org/10.1002/bdra.23093.

[20] David TJ, Illingworth CA. Diaphragmatic hernia in the south-west of England. J. Med. Genet. 1976 Aug 1;13(4):253-262.

[21] Kardon G, Ackerman KG, McCulley DJ, Shen Y, Wynn J, Shang L, et al. Congenital diaphragmatic hernias: From genes to mechanisms to therapies. DMM Dis Model Mech 2017;10:955-970. https://doi.org/10.1242/dmm.028365.

[22] Doray B, Girard-Lemaire F, Gasser B, Baldauf J-J, de Geeter B, Spizzo M, et al. Pallister-Killian syndrome: difficulties of prenatal diagnosis. Prenat Diagn 2002;22:470-477. https://doi.org/10.1002/pd.342.

[23] Frisova V, Svobodova IT, Tozzi M, Raskova D. Prenatal diagnosis of Pallister-Killian syndrome in pregnancy with normal CVS result and abnormal ultrasound findings in the second trimester. Taiwan J Obstet Gynecol 2018;57:726-729. https://doi.org/10.1016/j.tjog.2018.08.021.

[24] Kucińska-Chahwan A, Bijok J, Dąbkowska S, Jóźwiak A, Ilnicka A, Nowakowska B, et al. Targeted prenatal diagnosis of Pallister–Killian syndrome. Prenat Diagn 2017;37:446-452. https://doi.org/10.1002/pd.5030.

[25] Karaman B, Kayserili H, Ghanbari A, Uyguner ZO, Toksoy G, Altunoglu U, et al. Pallister-Killian syndrome: Clinical, cytogenetic and molecular findings in 15 cases. Mol Cytogenet 2018;11. https://doi.org/10.1186/s13039-018-0395-z.

[26] Salzano E, Raible S, Kaur M, Wilkens A, Sperti G, Tilton R, et al. Prenatal profile of Pallister-Killian syndrome: Retrospective analysis of 114 pregnancies, literature review and approach to prenatal diagnosis. Am J Med Genet Part A 2018;176:2575-2586. https://doi.org/10.1002/ajmg.a.40499.

[27] Wang T, Ren C, Chen D, Lu J, Guo L, Zheng L, et al. Prenatal diagnosis of Pallister-Killian syndrome using cord blood samples. Mol Cytogenet 2019;12:39. https://doi.org/10.1186/s13039-019-0449-x.

[28] Clarkson KB, Coming K, Toburen A. Lethal Pallister-Killian syndrome and Fryns syndrome: diagnostic considerations. Genet Med 2000;2:77-77. https://doi.org/10.1097/00125817-200001000-00091.

[29] Longoni M, Pober BR, High FA. Congenital Diaphragmatic Hernia Overview. InGeneReviews®[Internet] 2019 Mar 28. University of Washington, Seattle.

[30] Garne E, Haeusler M, Barisic I, Gjergja R, Stoll C, Clementi M. Congenital diaphragmatic hernia: evaluation of prenatal diagnosis in 20 European regions. Ultrasound Obstet Gynecol 2002;19:329-333. https://doi.org/10.1046/j.1469-0705.2002.00635.x

[31] Gallot D, Coste K, Francannet C, Laurichesse H, Boda C, Ughetto S, et al. Antenatal detection and impact on outcome of congenital diaphragmatic hernia: A 12-year experience in Auvergne (France). Eur J Obstet Gynecol Reprod Biol 2006;125:202-205. https://doi.org/10.1016/j.ejogrb.2005.06.030.

[32] Russo FM, Cordier A-G, De Catte L, Saada J, Benachi A, Deprest J. Proposal for standardized prenatal ultrasound assessment of the fetus with congenital diaphragmatic hernia by the European reference network on rare inherited and congenital anomalies (ERNICA). Prenat Diagn 2018;38:629-637. https://doi.org/10.1002/pd.5297.

[33] Metkus AP, Filly RA, Stringer MD, Harrison MR, Adzick NS. Sonographic predictors of survival in fetal diaphragmatic hernia. J. Pediatr. Surg., vol. 31, W.B. Saunders; 1996, p. 148-152. https://doi.org/10.1016/S0022-3468(96)90338-3.

[34] Peralta CFA, Cavoretto P, Csapo B, Vandecruys H, Nicolaides KH. Assessment of lung area in normal fetuses at 12-32 weeks. Ultrasound Obstet Gynecol 2005;26:718-724. https://doi.org/10.1002/uog.2651.

[35] Knox E, Lissauer D, Khan K, Kilby M. Prenatal detection of pulmonary hypoplasia in fetuses with congenital diaphragmatic hernia: A systematic review and meta-analysis of diagnostic studies. J Matern Neonatal Med 2010;23:579-588. https://doi.org/10.3109/14767050903551400.

[36] Odibo AO, Biron-Shental T, Tomlinson TM, Shim TL, Wanat K. Discussion: "Predictive value of lung-head ratio in congenital diaphragmatic hernia" by Yang et al. Am J Obstet Gynecol 2007;197:e1–e5. https://doi.org/10.1016/j.ajog.2007.05.050.

[37] Russo FM, Eastwood MP, Keijzer R, Al-Maary J, Toelen J, Van Mieghem T, et al. Lung size and liver herniation predict need for extracorporeal membrane oxygenation but not pulmonary hypertension in isolated congenital diaphragmatic hernia: systematic review and meta-analysis. Ultrasound Obstet Gynecol 2017;49:704-713. https://doi.org/10.1002/uog.16000.

[38] Jani J, Keller RL, Benachi A, Nicolaides KH, Favre R, Gratacos E, et al. Prenatal prediction of survival in isolated left-sided diaphragmatic hernia. Ultrasound Obstet Gynecol 2006;27:18-22. https://doi.org/10.1002/uog.2688.

[39] Lipshutz GS, Albanese CT, Feldstein VA, Jennings RW, Housley HT, Beech R, et al. Prospective analysis of lung-to-head ratio predicts survival for patients with prenatally diagnosed congenital diaphragmatic hernia.

J. Pediatr. Surg., vol. 32, W.B. Saunders; 1997, p. 1634-1636. https://doi.org/10.1016/S0022-3468(97)90471-1.

[40] Ruano R, Takashi E, Da Silva MM, Campos JADB, Tannuri U, Zugaib M. Prediction and probability of neonatal outcome in isolated congenital diaphragmatic hernia using multiple ultrasound parameters. Ultrasound Obstet Gynecol 2012;39:42-49. https://doi.org/10.1002/uog.10095.

[41] Jani J, Nicolaides KH, Keller RL, Benachi A, Peralta CFA, Favre R, et al. Observed to expected lung area to head circumference ratio in the prediction of survival in fetuses with isolated diaphragmatic hernia. Ultrasound Obstet Gynecol 2007;30:67-71. https://doi.org/10.1002/uog.4052.

[42] Deprest JA, Flemmer AW, Gratacos E, Nicolaides K. Antenatal prediction of lung volume and in-utero treatment by fetal endoscopic tracheal occlusion in severe isolated congenital diaphragmatic hernia. Semin Fetal Neonatal Med 2009;14:8-13. https://doi.org/10.1016/j.siny.2008.08.010.

[43] Snoek KG, Peters NCJ, van Rosmalen J, van Heijst AFJ, Eggink AJ, Sikkel E, et al. The validity of the observed-to-expected lung-to-head ratio in congenital diaphragmatic hernia in an era of standardized neonatal treatment; a multicenter study. Prenat Diagn. 2017;37(7):658-665.

[44] Russo FM, Cordier A-G, Salazar L, Basurto D, Gomez O, Gratacos E, et al. 203: Prediction of survival in right sided congenital diaphragmatic hernia in the FETO era. Am J Obstet Gynecol. 2019;220(1):S147. Available from: http://www.ajog.org/article/S0002937818312468/fulltext

[45] Alfaraj MA, Shah PS, Bohn D, Pantazi S, O'Brien K, Chiu PP, et al. Congenital diaphragmatic hernia: Lung-to-head ratio and lung volume for prediction of outcome. Am J Obstet Gynecol. 2011;205(1):43.e1-43.e8.

[46] Quintero RA, Kontopoulos E V., Quintero LF, Landy DC, Gonzalez R, Chmait RH. The observed vs. expected lung-to-head ratio does not correct for the effect of gestational age on the lung-to-head ratio. J. Matern. Neonatal Med. 2013;26(6):552-557.

[47] Ruano R, Aubry MC, Barthe B, Dumez Y, Benachi A. Three-dimensional ultrasonographic measurements of the fetal lungs for prediction of perinatal outcome in isolated congenital diaphragmatic hernia. J Obstet Gynaecol Res. 2009;35(6):1031-1041.

[48] Ruano R, Benachi A, Joubin L, Aubry M-C, Thalabard J-C, Dumez Y, et al. Three-dimensional ultrasonographic assessment of fetal lung volume as prognostic factor in isolated congenital diaphragmatic hernia. BJOG An Int J Obstet Gynaecol. 2004;111(5):423-429. http://doi.wiley.com/10.1111/j.1471-0528.2004.00100.x

[49] Ruano R, Aubry MC, Dumez Y, Zugaib M, Benachi A. Predicting neonatal deaths and pulmonary hypoplasia in isolated congenital diaphragmatic hernia using the sonographic fetal lung volume-body weight ratio. Am J Roentgenol. 2008;190(5):1216-1219.

[50] Mahieu-Caputo D, Sonigo P, Dommergues M, Fournet JC, Thalabard JC, Abarca C, et al. Fetal lung volume measurement by magnetic resonance imaging in congenital diaphragmatic hernia. BJOG An Int J Obstet Gynaecol. 2001;108(8):863-868. http://doi.wiley.com/10.1111/j.1471-0528.2001.00184.x

[51] Barnewolt CE, Kunisaki SM, Fauza DO, Nemes LP, Estroff JA, Jennings RW. Percent predicted lung volumes as measured on fetal magnetic

resonance imaging: a useful biometric parameter for risk stratification in congenital diaphragmatic hernia. J Pediatr Surg. 2007;42(1):193-197.

[52] Weidner M, Hagelstein C, Debus A, Walleyo A, Weiss C, Schoenberg SO, et al. MRI-Based Ratio of Fetal Lung Volume to Fetal Body Volume as a New Prognostic Marker in Congenital Diaphragmatic Hernia. Am J Roentgenol. 2014;202(6):1330-1336. http://www.ajronline.org/doi/abs/10.2214/AJR.13.11023

[53] Balassy C, Kasprian G, Brugger PC, Weber M, Csapo B, Herold C, et al. Assessment of lung development in isolated congenital diaphragmatic hernia using signal intensity ratios on fetal MR imaging. Eur Radiol. 2010;20(4):829-837. https://link.springer.com/article/10.1007/s00330-009-1633-x

[54] Nishie A, Tajima T, Asayama Y, Ishigami K, Hirakawa M, Nakayama T, et al. MR prediction of postnatal outcomes in left-sided congenital diaphragmatic hernia using right lung signal intensity: Comparison with that using right lung volume. J Magn Reson Imaging. 2009;30(1):112-120. http://doi.wiley.com/10.1002/jmri.21829

[55] Kilian AK, Schaible T, Hofmann V, Brade J, Neff KW, Büsing KA. Congenital diaphragmatic hernia: Predictive value of MRI relative lung-to-head ratio compared with MRI fetal lung volume and sonographic lung-to-head ratio. Am J Roentgenol. 2009;192(1):153-158.

[56] Vuletin JF, Lim FY, Cnota J, Kline-Fath B, Salisbury S, Haberman B, et al. Prenatal pulmonary hypertension index: novel prenatal predictor of severe postnatal pulmonary artery hypertension in antenatally diagnosed congenital diaphragmatic hernia. J Pediatr Surg. 2010;45(4):703-708.

[57] Cruz-Martinez R, Moreno-Alvarez O, Hernandez-Andrade E, Castañon M, Done E, Martinez JM, et al. Contribution of intrapulmonary artery Doppler to improve prediction of survival in fetuses with congenital diaphragmatic hernia treated with fetal endoscopic tracheal occlusion. Ultrasound Obstet Gynecol. 2010;35(5):572-577. Available from: http://doi.wiley.com/10.1002/uog.7593

[58] Cruz-Martinez R, Castañon M, Moreno-Alvarez O, Acosta-Rojas R, Martinez JM, Gratacos E. Usefulness of lung-to-head ratio and intrapulmonary arterial Doppler in predicting neonatal morbidity in fetuses with congenital diaphragmatic hernia treated with fetoscopic tracheal occlusion. Ultrasound Obstet Gynecol. 2013;41(1):59-65.

[59] Moreno-Alvarez O, Hernandez-Andrade E, Oros D, Jani J, Deprest J, Gratacos E. Association between intrapulmonary arterial Doppler parameters and degree of lung growth as measured by lung-to-head ratio in fetuses with congenital diaphragmatic hernia. Ultrasound Obstet Gynecol. 2008;31(2):164-170. Available from: http://doi.wiley.com/10.1002/uog.5201

[60] Sokol J, Shimizu N, Bohn D, Doherty D, Ryan G, Hornberger LK. Fetal pulmonary artery diameter measurements as a predictor of morbidity in antenatally diagnosed congenital diaphragmatic hernia: A prospective study. Am J Obstet Gynecol. 2006;195(2):470-477.

[61] Moreno-Alvarez O, Cruz-Martinez R, Hernandez-Andrade E, Done E, Gómez O, Deprest J, et al. Lung tissue perfusion in congenital diaphragmatic hernia and association with the lung-to-head ratio and intrapulmonary artery pulsed Doppler. Ultrasound Obstet Gynecol. 2010;35(5):578-582. http://doi.wiley.com/10.1002/uog.7592

[62] Done E, Allegaert K, Lewi P, Jani J, Gucciardo L, Van Mieghem T, et al. Maternal hyperoxygenation test in fetuses undergoing FETO for severe isolated congenital diaphragmatic hernia. Ultrasound Obstet Gynecol. 2011;37(3):264-271. Available from: http://doi.wiley.com/10.1002/uog.7753

[63] Ruano R, Aubry MC, Barthe B, Mitanchez D, Dumez Y, Benachi A. Predicting perinatal outcome in isolated congenital diaphragmatic hernia using fetal pulmonary artery diameters. J Pediatr Surg. 2008;43(4):606-611.

[64] Takahashi S, Oishi Y, Ito N, Nanba Y, Tsukamoto K, Nakamura T, et al. Evaluating mortality and disease severity in congenital diaphragmatic hernia using the McGoon and pulmonary artery indices. J Pediatr Surg. 2009;44(11):2101-2106.

[65] Kitano Y, Nakagawa S, Kuroda T, Honna T, Itoh Y, Nakamura T, et al. Liver position in fetal congenital diaphragmatic hernia retains a prognostic value in the era of lung-protective strategy. J Pediatr Surg. 2005;40(12):1827-1832.

[66] Werneck Britto IS, Olutoye OO, Cass DL, Zamora IJ, Lee TC, Cassady CI, et al. Quantification of liver herniation in fetuses with isolated congenital diaphragmatic hernia using two-dimensional ultrasonography. Ultrasound Obstet Gynecol. 2015;46(2):150-154. http://doi.wiley.com/10.1002/uog.14718

[67] Lazar DA, Ruano R, Cass DL, Moise KJ, Johnson A, Lee TC, et al. Defining "liver-up": Does the volume of liver herniation predict outcome for fetuses with isolated left-sided congenital diaphragmatic hernia? J Pediatr Surg; 2012;47(6):1058-1062. doi:10.1016/j.jpedsurg.2012.03.003.

[68] Cannie M, Jani J, Chaffiotte C, Vaast P, Deruelle P, Houfflin-Debarge V, et al. Quantification of intrathoracic liver herniation by magnetic resonance imaging and prediction of postnatal survival in fetuses with congenital diaphragmatic hernia. Ultrasound Obstet Gynecol. 2008;32(5):627-632.

[69] Kitano Y, Okuyama H, Saito M, Usui N, Morikawa N, Masumoto K, et al. Re-evaluation of stomach position as a simple prognostic factor in fetal left congenital diaphragmatic hernia: a multicenter survey in Japan. Ultrasound Obstet Gynecol. 2011;37(3):277-282. Available from: http://doi.wiley.com/10.1002/uog.8892

[70] Cordier AG, Cannie MM, Guilbaud L, De Laveaucoupet J, Martinovic J, Nowakowska D, et al. Stomach position versus liver-to-thoracic volume ratio in left-sided congenital diaphragmatic hernia. J Matern Neonatal Med. 2015;28(2):190-195.

[71] Basta AM, Lusk LA, Keller RL, Filly RA. Fetal Stomach Position Predicts Neonatal Outcomes in Isolated Left-Sided Congenital Diaphragmatic Hernia. Fetal Diagn Ther. 2016;39(4):248-255.

[72] Spaggiari E, Stirnemann J, Bernard J-P, De Saint Blanquat L, Beaudoin S, Ville Y. Prognostic value of a hernia sac in congenital diaphragmatic hernia. Ultrasound Obstet Gynecol. 2013;41(3):286-290. Available from: http://doi.wiley.com/10.1002/uog.11189

[73] Bouchghoul H, Marty O, Fouquet V, Cordier A-G, Senat M-V, Saada J, et al. Congenital diaphragmatic hernia has a better prognosis when associated with a hernia sac. Prenat Diagn. 2018;38(9):638-644. Available from: http://doi.wiley.com/10.1002/pd.5326

[74] Ruano R, Takashi E, Silva MM da, Haeri S, Tannuri U, Zugaib M. Quantitative Lung Index, Contralateral Lung Area, or Lung-to-Head

Ratio to Predict the Neonatal Outcome in Isolated Congenital Diaphragmatic Hernia? J Ultrasound Med. 2013;32(3):413-417. Available from: http://doi.wiley.com/10.7863/jum.2013.32.3.413

[75] DeKoninck P, Gomez O, Sandaite I, Richter J, Nawapun K, Eerdekens A, et al. Right-sided congenital diaphragmatic hernia in a decade of fetal surgery. BJOG An Int J Obstet Gynaecol. 2015;122(7):940-946. Available from: http://doi.wiley.com/10.1111/1471-0528.13065

[76] Kunisaki SM, Barnewolt CE, Estroff JA, Nemes LP, Jennings RW, Wilson JM, et al. Liver position is a prenatal predictive factor of prosthetic repair in congenital diaphragmatic hernia. Fetal Diagn Ther. 2008;23(4):258-262.

[77] Mullassery D, Ba'Ath ME, Jesudason EC, Losty PD. Value of liver herniation in prediction of outcome in fetal congenital diaphragmatic hernia: A systematic review and meta-analysis. Ultrasound Obstet Gynecol. 2010 May;35(5):609-614.

[78] Jani J, Cannie M, Sonigo P, Robert Y, Moreno O, Benachi A, et al. Value of prenatal magnetic resonance imaging in the prediction of postnatal outcome in fetuses with diaphragmatic hernia. Ultrasound Obstet Gynecol.2008 Nov 1;32(6):793-799. Available from: http://doi.wiley.com/10.1002/uog.6234

[79] Bebbington M, Victoria T, Danzer E, Moldenhauer J, Khalek N, Johnson M, et al. Comparison of ultrasound and magnetic resonance imaging parameters in predicting survival in isolated left-sided congenital diaphragmatic hernia. Ultrasound Obstet Gynecol. 2014;43(6):670-674. Available from: http://doi.wiley.com/10.1002/uog.13271

[80] Jani JC, Cannie M, Peralta CFA, Deprest JA, Nicolaides KH, Dymarkowski S. Lung volumes in fetuses with congenital diaphragmatic hernia: Comparison of 3D US and MR imaging assessments. Radiology. 2007;244(2):575-582.

[81] Puligandla PS, Skarsgard ED, Offringa M, Adatia I, Baird R, Bailey JAM, et al. Diagnosis and management of congenital diaphragmatic hernia: a clinical practice guideline. CMAJ. 2018;190(4):E103–E112.

[82] Lee TC, Lim FY, Keswani SG, Frischer JS, Haberman B, Kingma PS, et al. Late gestation fetal magnetic resonance imaging-derived total lung volume predicts postnatal survival and need for extracorporeal membrane oxygenation support in isolated congenital diaphragmatic hernia. J Pediatr Surg. 2011;46(6):1165-1171.

[83] Cannie M, Jani J, Meersschaert J, Allegaert K, Done' E, Marchal G, et al. Prenatal prediction of survival in isolated diaphragmatic hernia using observed to expected total fetal lung volume determined by magnetic resonance imaging based on either gestational age or fetal body volume. Ultrasound Obstet Gynecol. 2008 Oct 1;32(5):633-639. Available from: http://doi.wiley.com/10.1002/uog.6139

[84] Oluyomi-Obi T, Kuret V, Puligandla P, Lodha A, Lee-Robertson H, Lee K, Somerset D, Johnson J, Ryan G. Antenatal predictors of outcome in prenatally diagnosed congenital diaphragmatic hernia (CDH). J. Pediatr. Surg. 2017;52(5):881-888.

[85] Barnewolt CE, Kunisaki SM, Fauza DO, Nemes LP, Estroff JA, Jennings RW. Percent predicted lung volumes as measured on fetal magnetic resonance imaging: a useful biometric parameter for risk stratification in congenital diaphragmatic hernia. J. Pediatr. Surg. 2007;42(1):193-197.

[86] Shieh HF, Barnewolt CE, Wilson JM, Zurakowski D, Connolly SA, Estroff JA, et al. Percent predicted lung volume changes on fetal magnetic resonance imaging throughout gestation in congenital diaphragmatic hernia. J Pediatr Surg. 2017;52(6):933-937.

[87] Ruano R, Lazar DA, Cass DL, Zamora IJ, Lee TC, Cassady CI, Mehollin-Ray A, Welty S, Fernandes CJ, Haeri S, Belfort MA. Fetal lung volume and quantification of liver herniation by magnetic resonance imaging in isolated congenital diaphragmatic hernia. Ultrasound Obstet. Gynecol. 2014;43(6):662-669.

[88] Zamora IJ, Olutoye OO, Cass DL, Fallon SC, Lazar DA, Cassady CI, Mehollin-Ray AR, Welty SE, Ruano R, Belfort MA, Lee TC. Prenatal MRI fetal lung volumes and percent liver herniation predict pulmonary morbidity in congenital diaphragmatic hernia (CDH). J Pediatr Surg. 2014;49(5):688-693.

[89] Akinkuotu AC, Cruz SM, Abbas PI, Lee TC, Welty SE, Olutoye OO, Cassady CI, Mehollin-Ray AR, Ruano R, Belfort MA, Cass DL. Risk-stratification of severity for infants with CDH: prenatal versus postnatal predictors of outcome. J Pediatr Surg. 2016;51(1):44-48.

[90] Prayer F, Metzelder M, Krois W, Brugger PC, Gruber GM, Weber M, Scharrer A, Rokitansky A, Langs G, Prayer D, Unger E. Three-dimensional reconstruction of defects in congenital diaphragmatic hernia: a fetal MRI study. Ultrasound Obstet. Gynecol. 2019;53(6):816-826.

[91] Spaggiari E, Stirnemann JJ, Sonigo P, Khen-Dunlop N, De Saint Blanquat L, Ville Y. Prenatal prediction of pulmonary arterial hypertension in congenital diaphragmatic hernia. Ultrasound Obstet Gynecol. 2015;45(5):572-577.

[92] Done E, Debeer A, Gucciardo L, Van Mieghem T, Lewi P, Devlieger R, et al. Prediction of Neonatal Respiratory Function and Pulmonary Hypertension in Fetuses with Isolated Congenital Diaphragmatic Hernia in the Fetal Endoscopic Tracleal Occlusion Era: A Single-Center Study. Fetal Diagn Ther. 2015;37(1):24-32.

[93] Van Mieghem T, Cruz-Martinez R, Allegaert K, Dekoninck P, Castanon M, Sandaite I, et al. Outcome of fetuses with congenital diaphragmatic hernia and associated intrafetal fluid effusions managed in the era of fetal surgery. Ultrasound Obstet Gynecol. 2012;39(1):50-55. Available from: http://doi.wiley.com/10.1002/uog.10097

[94] Le LD, Keswani SG, Biesiada J, Lim FY, Kingma PS, Haberman BE, et al. The congenital diaphragmatic hernia composite prognostic index correlates with survival in left-sided congenital diaphragmatic hernia. J Pediatr Surg. 2012;47(1):57-62.

[95] Lally KP, Lally PA, Lasky RE, Tibboel D, Jaksic T, Wilson JM, et al. Defect size determines survival in infants with congenital diaphragmatic hernia. Pediatrics. 2007;120(3):e651–e657.

[96] Congenital diaphragmatic hernia: Defect size correlates with developmental defect. J Pediatr Surg. 2013;48:1177-82.

[97] Harting MT, Lally KP. The congenital diaphragmatic hernia study group registry update. Semin. Fetal Neonatal Med. 2014; 19(6):370-375.

[98] Lally KP, Lasky RE, Lally PA, Bagolan P, Davis CF, Frenckner BP, et al. Standardized reporting for congenital diaphragmatic hernia - An

international consensus. J Pediatr Surg. 2013;48(12):2408-2415. doi:10.1016/j.jpedsurg.2013.08.014.

[99] Schultz CM, DiGeronimo RJ, Yoder BA. Congenital diaphragmatic hernia: a simplified postnatal predictor of outcome. J Pediatr Surg. 2007;42(3):510-516.

[100] Lally KP, Jaksic T, Wilson JM, Clark RH, Hardin WD, Hirschl RB, et al. Estimating disease severity of congenital diaphragmatic hernia in the first 5 minutes of life. J Pediatr Surg. 2001;36(1):141-145.

[101] Kays DW, Talbert JL, Islam S, Larson SD, Taylor JA, Perkins J. Improved survival in left liver-up congenital diaphragmatic hernia by early repair before extracorporeal membrane oxygenation: optimization of patient selection by multivariate risk modeling. Journal of the American College of Surgeons. 2016;222(4):459-470.

[102] Gentili A, Pasini L, Iannella E, Landuzzi V, Lima M, Bacchi Reggiani ML, et al. Predictive outcome indexes in neonatal Congenital Diaphragmatic Hernia. J Matern Neonatal Med. 2015 ;28(13): 1602-1607.

[103] Jancelewicz T, Brindle ME, Harting MT, Tolley EA, Langham MR, Lally PA, et al. Extracorporeal Membrane Oxygenation (ECMO) Risk Stratification in Newborns with Congenital Diaphragmatic Hernia (CDH). J Pediatr Surg. 2018;53(10):1890-1895.

[104] Skarsgard ED, Macnab YC, Qiu Z, Little R, Lee SK. SNAP-II Predicts Mortality among Infants with Congenital Diaphragmatic Hernia. J Perinatol. 2005;25:315-319.

[105] Chiu LW, Desai J, Shanti C, Rane S, Agarwal P, Thomas RL, et al. SNAPPE II Score As a Predictor of Survival in Neonates with Congenital Diaphragmatic Hernia: A Single Center Experience. Eur J Pediatr Surg. 2016;26(4):316-321. Available from: http://www.thieme-connect.de/DOI/DOI?10.1055/s-0035-1554103

[106] Brindle ME, Cook EF, Tibboel D, Lally PA, Lally KP. A clinical prediction rule for the severity of congenital diaphragmatic hernias in newborns. Pediatrics. 2014;134(2):e413–e419.

[107] Bent DP, Nelson J, Kent DM, Jen HC. Population-based Validation of a Clinical Prediction Model for Congenital Diaphragmatic Hernias HHS Public Access. J Pediatr. 2018;201:160-165.

[108] Stoll C, Alembik Y, Dott B, Roth MP. Associated non diaphragmatic anomalies among cases with congenital diaphragmatic hernia. Genet Couns [Internet]. 2015;26(3):281-298.

[109] Montalva L, Lauriti G, Zani A. Congenital heart disease associated with congenital diaphragmatic hernia: A systematic review on incidence, prenatal diagnosis, management, and outcome. J Pediatr Surg. 2019;54(5):909-919.

[110] Schnitzer JJ, Hedrick HL, Pacheco BA, Losty PD, Ryan DP, Doody DP, Donahoe PK. Prenatal glucocorticoid therapy reverses pulmonary immaturity in congenital diaphragmatic hernia in fetal sheep. Annals of surgery. 1996;224(4):430.

[111] Schmidt AF, Gonçalves FLL, Nassr ACC, Pereira LAVD, Farmer D, Sbragia L. Antenatal steroid and tracheal occlusion restore vascular endothelial growth factor receptors in congenital diaphragmatic hernia rat model. Am J Obstet Gynecol. 2010;203(2):184.e13-184.e20.

[112] Suen HC, Bloch KD, Donahoe PK. Antenatal glucocorticoid corrects pulmonary immaturity in experimentally induced congenital diaphragmatic hernia in rats. Pediatr Res. 1994;35(5):523-529.

[113] Davey MG, Danzer E, Schwarz U, Adzick NS, Flake AW, Hedrick HL. Prenatal glucocorticoids and exogenous surfactant therapy improve respiratory function in lambs with severe diaphragmatic hernia following fetal tracheal occlusion. Pediatr Res. 2006;60(2):131-135.

[114] Mayer S, Klaritsch P, Sbragia L, Toelen J, Till H, Deprest JA. Maternal administration of betamethasone inhibits proliferation induced by fetal tracheal occlusion in the nitrofen rat model for congenital diaphragmatic hernia: a placebo-controlled study. Pediatric surgery international. 2008;24(12):1287-1295.

[115] Kay S, Flageole JM, Laberge H, Richardson S, Belanger S, Piedboeuf B. Use of antenatal steroids to counteract the negative effects of tracheal occlusion in the fetal lamb model. Pediatr Res. 2001;50(4):495-501.

[116] Davey MG, Danzer E, Schwarz U, Robinson L, Shegu S, Adzick NS, et al. Prenatal glucocorticoids improve lung morphology and partially restores surfactant mRNA expression in lambs with diaphragmatic hernia undergoing fetal tracheal occlusion. Pediatr Pulmonol. 2006;41(12):1188-1196. Available from: http://doi.wiley.com/10.1002/ppul.20516

[117] Roubliova XI, Lewi PJ, Verbeken EK, Vaast P, Jani JC, Lu H, et al. The effect of maternal betamethasone and fetal tracheal occlusion on pulmonary vascular morphometry in fetal rabbits with surgically induced diaphragmatic hernia: A placebo controlled morphologic study. Prenat Diagn. 2009;29(7):674-681.

[118] Lally KP, Bagolan P, Hosie S, Lally PA, Stewart M, Cotten CM, et al. Corticosteroids for fetuses with congenital diaphragmatic hernia: Can we show benefit? J Pediatr Surg. 2006;41(4):668-674.

[119] Gyamfi-Bannerman C, Thom EA, Blackwell SC, Tita ATN, Reddy UM, Saade GR, et al. Antenatal betamethasone for women at risk for late preterm delivery. N Engl J Med. 2016;374(14):1311-1320.

[120] Luong C, Rey-Perra J, Vadivel A, Gilmour G, Sauve Y, Koonen D, et al. Antenatal sildenafil treatment attenuates pulmonary hypertension in experimental congenital diaphragmatic hernia. Circulation. 2011;123(19):2120-2131.

[121] Kashyap AJ, Dekoninck PLJ, Rodgers KA, Thio M, Mcgillick EV, Amberg BJ, et al. Antenatal sildenafil treatment improves neonatal pulmonary hemodynamics and gas exchange in lambs with diaphragmatic hernia. Ultrasound Obstet Gynecol. 2019;54(4):506-516.

[122] Russo FM, Toelen J, Eastwood MP, Jimenez J, Miyague AH, Velde G Vande, et al. Transplacental sildenafil rescues lung abnormalities in the rabbit model of diaphragmatic hernia. Thorax. 2016 Jun 1;71(6):517-525.

[123] da Silva Ferreira RD, Negrini R, Bernardo WM, Simões R, Piato S. The effects of sildenafil in maternal and fetal outcomes in pregnancy: A systematic review and meta-analysis. PLoS One. 2019;14(7): e0219732.

[124] Sharp A, Cornforth C, Jackson R, Harrold J, Turner MA, Kenny LC, et al. Maternal sildenafil for severe fetal growth restriction (STRIDER): a multicentre, randomised,

placebo-controlled, double-blind trial. Lancet Child Adolesc Heal. 2018;2:93.

[125] Dunn L, Greer R, Flenady V, Kumar S. Sildenafil in pregnancy: a systematic review of maternal tolerance and obstetric and perinatal outcomes. Fetal Diagn. Ther. 2017;41:81-88. https://doi.org/10.1159/000453062

[126] Russo FM, Benachi A, Van Mieghem T, De Hoon J, Van Calsteren K, Annaert P, et al. Antenatal sildenafil administration to prevent pulmonary hypertension in congenital diaphragmatic hernia (SToP-PH): study protocol for a phase I/IIb placenta transfer and safety study. Trials. 2018;19:524. https://doi.org/10.1186/s13063-018-2897-8.

[127] Schmidt AF, Gonçalves FLL, Regis AC, Gallindo RM, Sbragia L. Prenatal retinoic acid improves lung vascularization and VEGF expression in CDH rat. Am J Obstet Gynecol. 2012;207(1):76.e25-76.e32.

[128] Burgos CM, Davey MG, Riley JS, Jia H, Flake AW, Peranteau WH. Lung function and pulmonary artery blood flow following prenatal maternal retinoic acid and imatinib in the nitrofen model of congenital diaphragmatic hernia. J Pediatr Surg. 2018;53(9):1681-1687.

[129] Islam S, Narra V, Coté GM, Manganaro TF, Donahoe PK, Schnitzer JJ. Prenatal vitamin E treatment improves lung growth in fetal rats with congenital diaphragmatic hernia. J Pediatr Surg. 1999;34(1):172-177.

[130] Sakai K, Kimura O, Furukawa T, Fumino S, Higuchi K, Wakao J, et al. Prenatal administration of neuropeptide bombesin promotes lung development in a rat model of nitrofen-induced congenital diaphragmatic hernia. J Pediatr Surg. 2014;49(12):1749-1752.

[131] Takayama S, Sakai K, Fumino S, Furukawa T, Kishida T, Mazda O, et al. An intra-amniotic injection of mesenchymal stem cells promotes lung maturity in a rat congenital diaphragmatic hernia model. Pediatr Surg Int. 2019;35(12):1353-1361.

[132] Grushka JR, Laberge JM, Puligandla P, Skarsgard ED. Effect of hospital case volume on outcome in congenital diaphragmatic hernia: the experience of the Canadian Pediatric Surgery Network. J Pediatr Surg. 2009;44(5):873-876.

[133] Morche J, Mathes T, Jacobs A, Pietsch B, Wessel L, Gruber S, Neugebauer EA, Pieper D. Relationship between volume and outcome for surgery on congenital diaphragmatic hernia: a systematic review. J Pediatr Surg. https://doi.org/10.1016/j.jpedsurg.2020.03.025

[134] Nasr A, Langer JC. Influence of location of delivery on outcome in neonates with congenital diaphragmatic hernia. J Pediatr Surg. 2011;46(5):814-816.

[135] Carmichael SL, Ma C, Lee HC, Shaw GM, Sylvester KG, Hintz SR. Survival of infants with congenital diaphragmatic hernia in California: impact of hospital, clinical, and sociodemographic factors. J Perinatol. 2020;40(6):943-951. https://doi.org/10.1038/s41372-020-0612-6

[136] Snoek KG, Capolupo I, Van Rosmalen J, De Jongste-Van Den Hout L, Vijfhuize S, Greenough A, et al. Conventional mechanical ventilation versus high-frequency oscillatory ventilation for congenital diaphragmatic hernia. A randomized clinical trial (The VICI-trial). Ann Surg. 2016;263(5):867-874.

[137] Burgos CM, Frenckner B, Luco M, Harting MT, Lally PA, Lally KP. Prenatally diagnosed

congenital diaphragmatic hernia: Optimal mode of delivery? J Perinatol. 2017;37(2):134-138.

[138] Hutcheon J, Butler B, Lisonkova S, Marquette G, Mayer C, Skoll A, et al. Timing of delivery for pregnancies with congenital diaphragmatic hernia. BJOG An Int J Obstet Gynaecol. 2010;117(13):1658-1662. Available from: http://doi.wiley.com/10.1111/j.1471-0528.2010.02738.x

[139] Safavi A, Lin Y, Skarsgard ED. Perinatal management of congenital diaphragmatic hernia: When and how should babies be delivered? Results from the Canadian Pediatric Surgery Network. J Pediatr Surg. 2010;45(12):2334-2339.

[140] Ali K, Grigoratos D, Cornelius V, Nicolaides K, Davenport M, Greenough A. Gestation at delivery and outcome in CDH infants following fetoscopic tracheal occlusion. Eur Respir J. 2012;40(Suppl 56).

[141] Danzer E, Hoffman C, D'Agostino JA, Gerdes M, Bernbaum J, Antiel RM, et al. Neurodevelopmental outcomes at 5 years of age in congenital diaphragmatic hernia. J Pediatr Surg. 2017;52(3):437-443.

[142] Wyckoff MH, Aziz K, Escobedo MB, Kapadia VS, Kattwinkel J, Perlman JM, et al. Part 13: Neonatal resuscitation: 2015 American Heart Association guidelines update for cardiopulmonary resuscitation and emergency cardiovascular care. Circulation. 2015;132(18):S543–S560.

[143] Snoek KG, Reiss IKM, Greenough A, Capolupo I, Urlesberger B, Wessel L,6et al. Standardized Postnatal Management of Infants with Congenital Diaphragmatic Hernia in Europe: The CDH EURO Consortium Consensus - 2015 Update. Neonatology. 2016;110(1):66-74.

[144] Snoek KG. A Dive into the Wondrous World of Congenital Diaphragmatic Hernia: An international multicenter clinical approach [Thesis]. Erasmus MC: University Medical Center Rotterdam; 2016.

[145] Foglia EE, Ades A, Hedrick HL, Rintoul N, Munson DA, Moldenhauer J, et al. Initiating resuscitation before umbilical cord clamping in infants with congenital diaphragmatic hernia: A pilot feasibility trial. Arch Dis Child Fetal Neonatal Ed. 2020 May 1;105(3):F322–F326.

[146] Boucherat O, Benachi A, Chailley-Heu B, Franco-Montoya ML, Elie C, Martinovic J, et al. Surfactant maturation is not delayed in human fetuses with diaphragmatic hernia. PLoS Med. 2007;4(7):1205-1215.

[147] Cogo PE, Simonato M, Danhaive O, Verlato G, Cobellis G, Savignoni F, et al. Impaired surfactant protein B synthesis in infants with congenital diaphragmatic hernia. Eur Respir J. 2013;41(3):677-682.

[148] Van Meurs K. Is surfactant therapy beneficial in the treatment of the term newborn infant with congenital diaphragmatic hernia? J Pediatr. 2004;145(3):312-316.

[149] Lally KP, Lally PA, Langham MR, Hirschl R, Moya FR, Tibboel D, et al. Surfactant does not improve survival rate in preterm infants with congenital diaphragmatic hernia. J Pediatr Surg. 2004;39(6):829-833.

[150] Colby CE. Surfactant replacement therapy on ECMO does not improve outcome in neonates with congenital diaphragmatic hernia. J Pediatr Surg. 2004;39(11):1632-1637.

[151] Logan JW, Rice HE, Goldberg RN, Cotten CM. Congenital diaphragmatic

hernia: a systematic review and summary of best-evidence practice strategies. J Perinatol. 2007;27(9):535-549.

[152] Hagadorn JI, Brownell EA, Herbst KW, Trzaski JM, Neff S, Campbell BT. Trends in treatment and in-hospital mortality for neonates with congenital diaphragmatic hernia. J Perinatol. 2015;35:748-754.

[153] Guidry CA, Hranjec T, Rodgers BM, Kane B, McGahren ED. Permissive hypercapnia in the management of congenital diaphragmatic hernia: our institutional experience.J Am Coll Surg. 2012; 214:640-47. doi: 10.1016/j.jamcollsurg.2011.12.036.

[154] Puligandla PS, Grabowski J, Austin M, Hedrick H, Renaud E, Arnold M, et al. Management of congenital diaphragmatic hernia: A systematic review from the APSA outcomes and evidence based practice committee. Vol. 50, J Pediatr Surg. 2015;50:1958-1970.

[155] Wung JT, Sahni R, Moffitt ST, Lipsitz E, Stolar CJH. Congenital diaphragmatic hernia: Survival treated with very delayed surgery, spontaneous respiration, and no chest tube. J Pediatr Surg. 1995;30(3):406-409.

[156] Boloker J, Bateman DA, Wung JT, Stolar CJH. Congenital diaphragmatic hernia in 120 infants treated consecutively with permissive hypercapnea/spontaneous respiration/elective repair. J Pediatr Surg. 2002;37(3):357-366.

[157] Finer NN, Tierney A, Etches PC, Peliowski A, Ainsworth W. Congenital diaphragmatic hernia: Developing a protocolized approach. J Pediatr Surg. 1998;33(9):1331-1337.

[158] Bagolan P, Casaccia G, Crescenzi F, Nahom A, Trucchi A, Giorlandino C, et al. Impact of a Current Treatment Protocol on Outcome of High-Risk Congenital Diaphragmatic Hernia. J Pediatr Surg; 2004;39(3):313-318.

[159] Murthy V, D'Costa W, Nicolaides K, Davenport M, Fox G, Milner AD, et al. Neuromuscular Blockade and Lung Function during Resuscitation of Infants with Congenital Diaphragmatic Hernia. Neonatology. 2013;103(2):112-117.

[160] Harris J, Ramelet AS, van Dijk M, Pokorna P, Wielenga J, Tume L, et al. Clinical recommendations for pain, sedation, withdrawal and delirium assessment in critically ill infants and children: an ESPNIC position statement for healthcare professionals. Intensive care medicine. 2016;42(6):972-986.

[161] Giordano V, Edobor J, Deindl P, Wildner B, Goeral K, Steinbauer P, et al. Pain and Sedation Scales for Neonatal and Pediatric Patients in a Preverbal Stage of Development: A Systematic Review. JAMA Pediatrics. 2019;173(12):1186-1197.

[162] Watterberg KL, Cummings JJ, Benitz WE, Eichenwald EC, Poindexter BB, Stewart DL et al. Prevention and management of procedural pain in the neonate: An update. Pediatrics. 2016;137(2). e20154271. https://doi.org/10.1542/peds.2015-4271

[163] Patel N, Lally PA, Kipfmueller F, Massolo AC, Luco M, Van Meurs KP, et al. Ventricular dysfunction is a critical determinant of mortality in congenital diaphragmatic hernia. Am J Respir Crit Care Med. 2019 Dec 15;200(12):1522-1530.

[164] Byrne FA, Keller RL, Meadows J, Miniati D, Brook MM, Silverman NH, et al. Severe left diaphragmatic hernia limits size of fetal left heart more than does right diaphragmatic

hernia. Ultrasound Obstet Gynecol. 2015;46(6):688-694.

[165] Buijs EAB, Reiss IKM, Kraemer U, Andrinopoulou ER, Zwiers AJM, Ince C, et al. Increasing mean arterial blood pressure and heart rate with catecholaminergic drugs does not improve the microcirculation in children with congenital diaphragmatic hernia: A prospective Cohort study. Pediatr Crit Care Med. 2014;15(4):343-354.

[166] Patel N. Use of milrinone to treat cardiac dysfunction in infants with pulmonary hypertension secondary to congenital diaphragmatic hernia: a review of six patients. Neonatology. 2012;102(2):130-136.

[167] Sluiter I, van der Horst I, van der Voorn P, Boerema-de Munck A, Buscop-van Kempen M, de Krijger R, et al. Premature differentiation of vascular smooth muscle cells in human congenital diaphragmatic hernia. Exp Mol Pathol. 2013;94(1):195-202.

[168] Mous DS, Buscop-van Kempen MJ, Wijnen RMH, Tibboel D, Rottier RJ. Changes in vasoactive pathways in congenital diaphragmatic hernia associated pulmonary hypertension explain unresponsiveness to pharmacotherapy. Respir Res. 2017; 18(1);187.

[169] Solari V, Piaseczna Piotrowska A, Puri P. Expression of heme oxygenase-1 and endothelial nitric oxide synthase in the lung of newborns with Congenital diaphragmatic hernia and persistent pulmonary hypertension. J. Pediatr. Surg. 2003; 38(5):808-813.

[170] Sood BG, Wykes S, Landa M, De Jesus L, Rabah R. Expression of eNOS in the lungs of neonates with pulmonary hypertension. Exp Mol Pathol. 2011;90(1):9-12.

[171] Shinkai T, Shima H, Solari V, Puri P. Expression of vasoactive mediators during mechanical ventilation in nitrofen-induced diaphragmatic hernia in rats. Pediatr Surg Int; 2005;21(3):143-147.

[172] Shehata SMK, Sharma HS, Mooi WJ, Tibboel D. Pulmonary hypertension in human newborns with congenital diaphragmatic hernia is associated with decreased vascular expression of nitric-oxide synthase. Cell Biochem Biophys. 2006;44(1):147-155.

[173] Montalva L, Antounians L, Zani A. Pulmonary hypertension secondary to congenital diaphragmatic hernia: factors and pathways involved in pulmonary vascular remodeling. Pediatric research. 2019 ;85(6):754-768.

[174] Nair J, Lakshminrusimha S. Update on PPHN: Mechanisms and treatment. Semin. Perinatol. 2014;38(2):78-91.

[175] Keller RL, Tacy TA, Hendricks-MunozK,XuJ,Moon-GradyAJ, NeuhausJ,etal.Congenitaldiaphragmatic hernia: Endothelin-1, pulmonary hypertension, and disease severity. Am J Respir Crit Care Med. 2010;182(4):555-561.

[176] Mohseni-Bod H, Bohn D. Pulmonary hypertension in congenital diaphragmatic hernia. Semin Pediatr Surg. 2007 ;16(2):126-133.

[177] Hofmann SR, Stadler K, Heilmann A, Häusler HJ, Fitze G, Kamin G, et al. Stabilisation of cardiopulmonary function in newborns with congenital diaphragmatic hernia using lung function parameters and hemodynamic management. Klin Padiatr. 2012;224(4).

[178] Gien J, Kinsella JP. Management of pulmonary hypertension in infants with congenital diaphragmatic hernia. J. Perinatol. 2016;36(2):S28-S31.

[179] Kumar VH., Dadiz R, Koumoundouros J, Lakshminrusimha S.

Response to Pulmonary Vasodilators in Infants with Congenital Diaphragmatic Hernia. Pediatrics. 2018;142:184-184.

[180] McNamara PJ, Laique F, Muang-In S, Whyte HE. Milrinone improves oxygenation in neonates with severe persistent pulmonary hypertension of the newborn. J Crit Care. 2006;21(2):217-222.

[181] Bassler D, Choong K, McNamara P, Kirpalani H. Neonatal persistent pulmonary hypertension treated with milrinone: Four case reports. Biol Neonate. 2006;89(1):1-5.

[182] Lakshminrusimha S, Keszler M, Kirpalani H, Van Meurs K, Chess P, Ambalavanan N, et al. Milrinone in congenital diaphragmatic hernia – a randomized pilot trial: study protocol, review of literature and survey of current practices. Matern Heal Neonatol Perinatol. 2017;3(1):1-5.

[183] Finer NM. Inhaled nitric oxide and hypoxic respiratory failure in infants with congenital diaphragmatic hernia. Pediatrics. 1997;99(6):838-845.

[184] Putnam LR, Tsao K, Morini F, Lally PA, Miller CC, Lally KP, et al. Evaluation of variability in inhaled nitric oxide use and pulmonary hypertension in patients with congenital diaphragmatic hernia. JAMA Pediatr. 2016;170(12):1188-1194.

[185] Kelly LE, Ohlsson A, Shah PS. Sildenafil for pulmonary hypertension in neonates. Cochrane Database of Systematic Reviews. 2017(8). https://doi.org/10.1002/14651858.CD005494.pub4

[186] Noori S, Friedlich P, Wong P, Garingo A, Seri I. Cardiovascular effects of sildenafil in neonates and infants with congenital diaphragmatic hernia and pulmonary hypertension. Neonatology. 2007;91(2):92-100. https://doi.org/10.1159/000097125

[187] Bialkowski A, Moenkemeyer F, Patel N. Intravenous sildenafil in the management of pulmonary hypertension associated with congenital diaphragmatic hernia. Eur J Pediatr Surg. 2015; 25(2):171-176. https://doi.org/10.1055/s-0033-1357757.

[188] Kipfmueller F, Schroeder L, Berg C, Heindel K, Bartmann P, Mueller A. Continuous intravenous sildenafil as an early treatment in neonates with congenital diaphragmatic hernia. Pediatr Pulmonol 2018;53:452-460. https://doi.org/10.1002/ppul.23935.

[189] Cochius-den Otter S, Schaible T, Greenough A, van Heijst A, Patel N, Allegaert K, et al. The CoDiNOS trial protocol: an international randomised controlled trial of intravenous sildenafil versus inhaled nitric oxide for the treatment of pulmonary hypertension in neonates with congenital diaphragmatic hernia. BMJ open. 2019;9:e032122. https://doi.org/10.1136/bmjopen-2019-032122.

[190] More K, Athalye-Jape GK, Rao SC, Patole SK. Endothelin receptor antagonists for persistent pulmonary hypertension in term and late preterm infants. Cochrane Database of Systematic Reviews. 2016(8). https://www.cochranelibrary.com/cdsr/doi/10.1002/14651858.CD010531.pub2.

[191] Steinhorn RH, Fineman J, Kusic-Pajic A, Cornelisse P, Gehin M, Nowbakht P, et al. Bosentan as Adjunctive Therapy for Persistent Pulmonary Hypertension of the Newborn: Results of the Randomized Multicenter Placebo-Controlled Exploratory Trial. J Pediatr. 2016;177:90-96.e3. https://doi.org/10.1016/j.jpeds.2016.06.078.

[192] Pedersen J, Hedegaard ER, Simonsen U, Krüger M, Infanger M, Grimm D. Current and Future Treatments for Persistent Pulmonary Hypertension in the

Newborn. Basic Clin Pharmacol Toxicol. 2018;123(4):392-406. http://doi.wiley.com/10.1111/bcpt.13051

[193] Turek JW, Nellis JR, Sherwood BG, Kotagal M, Mesher AL, Thiagarajan RR, et al. Shifting Risks and Conflicting Outcomes—ECMO for Neonates with Congenital Diaphragmatic Hernia in the Modern Era. J Pediatr. 2017;190:163-168.e4. https://doi.org/10.1016/j.jpeds.2017.08.010.

[194] Saugstad OD. When increased mortality indicates improved care: CDH ECMO registry data. J Pediatr. 2017;190:4-5. https://doi.org/10.1016/j.jpeds.2017.09.037.

[195] Kays DW. ECMO in CDH: Is there a role? Semin Pediatr Surg. 2017;26(3):166-170. https://doi.org/10.1053/j.sempedsurg.2017.04.006.

[196] Section on Surgery and the Committee on Fetus and Newborn. Postdischarge follow-up of infants with congenital diaphragmatic hernia. Pediatrics. 2008 ;121(3):627-632.

[197] IJsselstijn H, Breatnach C, Hoskote A, Greenough A, Patel N, Capolupo I, et al. Defining outcomes following congenital diaphragmatic hernia using standardised clinical assessment and management plan (SCAMP) methodology within the CDH EURO consortium. Pediatr Res 2018;84:181-189. https://doi.org/10.1038/s41390-018-0063-3.

[198] Poley MJ, Stolk EA, Tibboel D, Molenaar JC, Busschbach JJV. Short term and long term health related quality of life after congenital anorectal malformations and congenital diaphragmatic hernia. Arch Dis Child 2004;89:836-841. https://doi.org/10.1136/adc.2002.016543.

[199] Fritz KA, Khmour AY, Kitzerow K, Sato TT, Basir MA. Health-related quality of life, educational and family outcomes in survivors of congenital diaphragmatic hernia. Pediatr Surg Int. 201935(3):315-20. https://doi.org/10.1007/s00383-018-4414-2.

[200] Michel F, Baumstarck K, Gosselin A, Le Coz P, Merrot T, Hassid S, et al. Health-related quality of life and its determinants in children with a congenital diaphragmatic hernia. Orphanet J Rare Dis. 2013;8(1). https://doi.org/10.1186/1750-1172-8-89.

[201] Bojanić K, Grizelj R, Vuković J, Omerza L, Grubić M, Ćaleta T, et al. Health-related quality of life in children and adolescents with congenital diaphragmatic hernia: A cross-sectional study. Health Qual Life Outcomes. 2018;16(1). https://doi.org/10.1186/s12955-018-0869-z.

Chapter 8

Transposition of Great Arteries

Rita Prasad Verma

Abstract

TGA is the commonest complex congenital cyanotic cardiac anomaly occurring during the first week of life. It is characterized by the unusual anomaly of ventriculoarterial discordance, with the aorta (A) originating from the right ventricle (RV) and the pulmonary artery (PA) from the left ventricle (LV). In the common Dextro form (DTGA), A is abnormally located to the right, anterior, and inferior of PA. The anatomic configuration results in the lethal hemodynamic pattern of 2 independent and parallel running circulatory circuits, which mandates creating a conduit to ensure the mixing of oxygenated and deoxygenated blood for survival. In the rare Levo form (LTGA), the aorta is placed anterior and to the left of PA with ventricular inversion. TGA is well tolerated in the fetus and is challenging to diagnose by fetal echocardiography unless the outflow tracts are specifically visualized. Postnatally the typical findings of murmur and cyanosis vary according to the associated cardiac defects and the degree of intercirculatory mixing. The arterial switch operation (ASO), which involves establishing ventriculoarterial concordance, is the standard surgical repair of D-TGA and has replaced the atrial switch procedures due to its superior long-term outcomes. The Rastelli procedure is used for complex DTGA cases. DTGA has a 90% mortality rate in the first year of life if untreated, while over 95% survive for 5 to 25 years after surgery. Post-surgical course may be complicated and require surgical revisions. The long-term outcome is associated with normal or mild to moderate neurodevelopmental disabilities, depending upon the type, complexity, and course of the disease. Expert follow-up of the patients into adulthood is an integral part of the management of TGA for best outcomes.

Keywords: Dextro-transposition of great arteries, Ventriculoarterial discordance, Ventriculoatrial discordance, Levo-transposition of great arteries, Arterial switch operation, Atrial switch operation

1. Introduction

Transposition of the great arteries (TGA) is a critical complex congenital cyanotic cardiac disorder, characterized by the unique anomaly of ventriculoarterial discordance of major vessels. In the common type, the aorta (Ao) arises abnormally from the right ventricle (RV) and the pulmonary artery (PA) from the left ventricle (LV), leading to a lethal hemodynamic pattern of 2 independent circulations, running parallel to each other. This condition mandates a conduit that will ensure the mixing of blood between the two vascular circuits for survival. An open ductus arteriosus (DA) and, to some extent, patent foramen ovale (PFO) serve this purpose during the initial hours of postnatal life. As the DA closes physiologically after birth, the condition becomes critical and requires the emergent reopening of the ductus or creating an atrial shunt to maintain the intercirculatory mixing of blood. TGA can result in acute cardiorespiratory decompensation and death within the first 48 hours of life if the diagnosis is missed. Early palliation followed by surgical repair, which involves physiological correction with the atrial switch or anatomic correction with arterial switch, has dramatically increased the survival rate of TGA to more than 90% from a universally fatal disease.

TGA is classified as Dextro (D-TGA) and Levo transpositions (L-TGA), based on Ao's relationship with PA in the anomalous heart. As LTGA is extremely rare, DTGA is discussed here, with the former referred to for comparison.

2. Epidemiology

TGA is the second most common congenital cyanotic heart defect (CCHD) and the commonest one occurring during the 1st week of life [1]. According to an estimation, about 1153 infants are born with TGA annually in the USA. The prevalence is assessed to be 2.3–4.7 per 10,000 live births [2, 3]. TGA accounts for approximately 3 percent of all congenital heart disease (CHD) disorders and almost 20 percent of all cyanotic CHD defects [2]. Overall, the incidence of DTGA is 5–10% of all CHDs, whereas that of LTGA is <1%, and 0.02 to 0.07 per 1000 live births. About 90% of the cases present as an isolated defect and the disorder is rarely associated with extracardiac anomalies. The occurrence of non-cardiac congenital lesions in TGA at <10% is significantly lower than those in other CHDs [4]. DTGA is not associated with any identifiable syndromes or genetic abnormality. It is notable that 80 percent of the patients with DiGeorge syndrome display 22q11 deletion and conotruncal lesions, but they rarely suffer from DTGA [5]. TGA in family members is uncommon, and the prevalence of CHD in the siblings of affected babies is not different from the general population at 0.3 percent [6]. TGA has a documented association with maternal diabetes mellitus, and it is more common in males to females in a 3:1 ratio in the DTGA form. The pathogenesis is multifactorial, and a combination of genetic and environmental factors is believed to play a role.

3. Embryology

While the exact embryology is undefined, TGA is hypothesized to be secondary to a developmental aberration in the morphogenesis of the bilateral sub arterial conus. During the first month of fetal life, the subaortic conus and sub pulmonary conus, which represent the preliminary great arteries, are normally positioned above the right ventricle. At approximately 30 to 34 days of fetal life, the subaortic CONUS is resorbed, and the aortic valve migrates inferiorly and posteriorly into the left ventricle. The sub pulmonary conus does not resorb, and the pulmonary valve retains its association with the right ventricle [7]. In D-TGA, the sub pulmonary CONUS is resorbed abnormally, and the pulmonary valve migrates posteriorly. Simultaneously, the unresorbed subaortic conus forces the aortic valve to move anteriorly and get engaged with the morphologic right ventricle. The origin and course of the coronary arteries vary and are determined by the movement of the subaortic conus [8].

4. Anatomy

Under normal conditions, Ao originates from the LV and is situated posteroinferior and to the right of the main pulmonary artery (**Figure 1**). In DTGA, Ao arises from the right ventricle (RV) and is positioned anteroinferior and to the right of PA, which is connected to the LV (**Figure 2**) [7]. The atria are normally positioned with atrial situs solitus, which is associated with atrioventricular concordance, d-looping of the ventricles, and ventriculo-arterial discordance. The aortic valve is anteroinferior and to the pulmonary valve's right instead of being posteroinferior and right as in normal conditions. The pulmonary and systemic circulations run independent of and parallel to each other with no mixing of the oxygenated and deoxygenated blood. In LTGA, the Ao is anterior, and to the left of PA, and the ventricles are inverted with atrioventricular discordance. The morphological RV is positioned to the left and the morphological LV to the right with Ao originating from the RV and PA from LV (**Figure 3**). The blood is pumped into the systemic circuit via RV and the pulmonary circuit via LV in this form. The systemic and pulmonary circulations are not impaired, and a shunt is not needed to mix blood. The two types of TGA are, therefore, hemodynamically different entities. In DTGA, with complete transposition and atrioventricular concordance, the lifesaving communication between the two parallel circulations is achieved interarterially via PDA, which connects the Ao with PA, or intra atrially via atrial septal defect (ASD) or PFO. Ventricular septal defect (VSD) is present in about 50% of the cases and provides another source of mixing between the oxygenated blood in the pulmonary circuit and deoxygenated blood in the systemic circuit.

Anatomically, DTGA may occur as an isolated defect (simple) or in combination with other cardiac lesions (complex). The commonest anomaly found in DTGA is VSD, often associated with other cardiac lesions, such as pulmonic valve stenosis

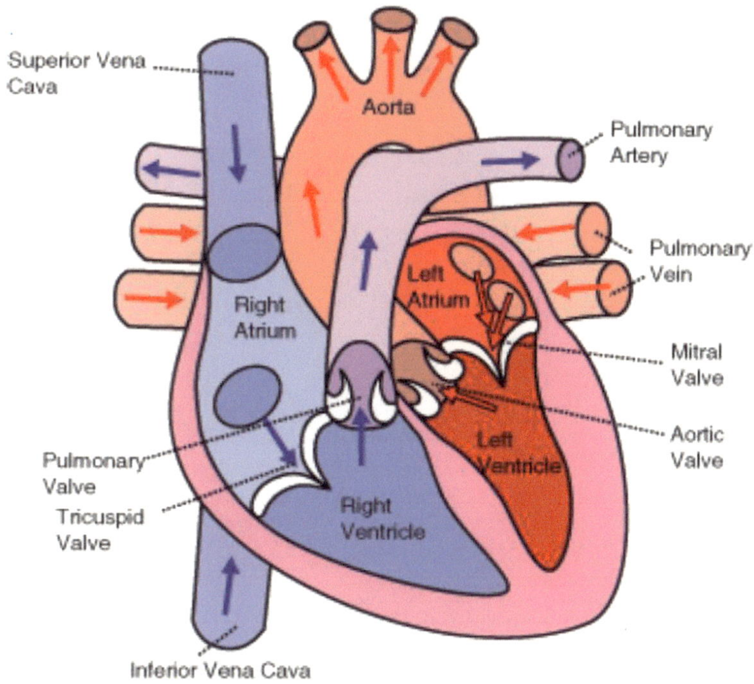

Figure 1.
Normal heart.

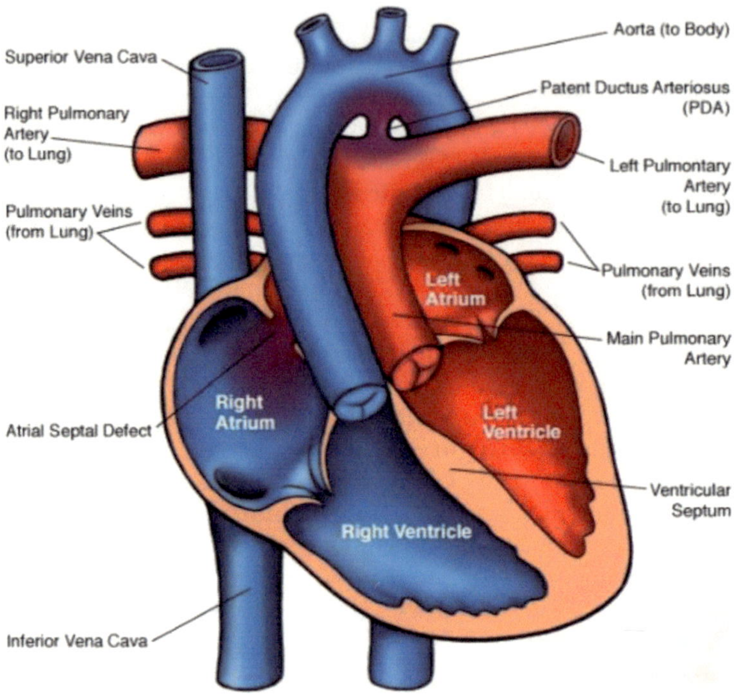

Figure 2.
Anatomy of heart in DTGA.

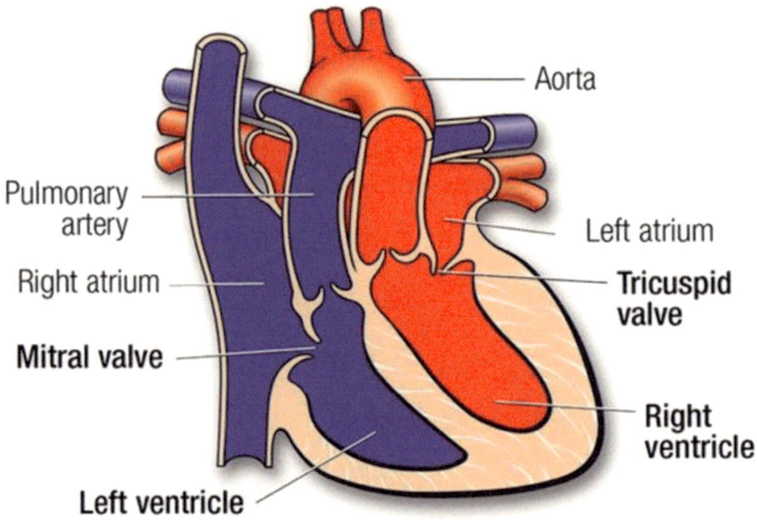

Figure 3.
Anatomy of heart in LTGA.

(PS) or atresia, overriding atrioventricular valves, coarctation of the aorta (COA), and aortic interruption [6–8]. The second most common defect in DTGA is left ventricular outflow tract obstruction (LVOO), detected in approximately 25% of the cases [9, 10]. LVOO may be anatomic or dynamic. If the interventricular septum is intact, the high pressure in RV, which may be at times equal to the systemic blood pressure, forces the ventricular septum to bulge into the LV cavity and thus creates a dynamic outflow obstruction. The dynamic obstruction is resolved when a conduit is formed between the two parallel circulations, either medically by reopening the DA with PGE1 or surgically by creating an atrial septal defect (ASD) by balloon septostomy. Other cardiac structural anomalies seen in DTGA involve atrioventricular valves, such as overriding valves or the straddling tricuspid valve [11].

In LTGA the defect is congenitally corrected with atrioventricular discordance. Ventricles are inverted, and A arises from RV and PA from LV [7]. Despite transposition of the great arteries, deoxygenated blood is pumped into the lungs, while oxygenated blood circulates to the rest of the body. With this anomaly, therefore, a shunt is not needed. Anatomically, the aortic valve is anterior and to the left of the pulmonary valve. LTGA is associated more often with other cardiac anomalies, the common ones being VSD at 60–70%, PS at 30–50%, tricuspid regurgitation (TR) at 30% and dextrocardia at <1%.

The coronary arteries are anatomically abnormal in about 33% of the patients with TGA [12]. The commonly reported anomalies are the left circumflex coronary originating from the right coronary artery in 22%, single right artery in 9.5%, single left coronary artery in 3%, or inverted coronary arteries in 3% of the cases. Their course towards their destination may be shortened and unusual, such as passing between the two great arteries. There may be multiple coronary ostia arising from the sinus of Valsalva [12].

5. Hemodynamics and pathophysiology

In DTGA, the deoxygenated blood coming from the superior and inferior vena cava normally drain into the right atrium (RA). From there, it passes over to the right ventricle (RV), then enters the systemic circulation via the abnormally connected Ao, and finally returns to the RA through the vena cava (**Figure 4**). The oxygenated blood in the pulmonary circuit enters the left ventricle (LV) via the left atrium (LA), is pumped into the abnormally connected pulmonary artery, and returns back to the LA via pulmonary veins. This circulatory pattern is incompatible with life and requires mandatory mixing between the two parallel circulations, which is achieved intracardiacally via the PFO, ASD or a VSD; or extracardiacally with a patent DA or other vascular channel of the bronchopulmonary collateral circulation.

DTGA is well tolerated by the fetus as the intercirculatory mixing of blood is maintained by the open fosa ovalis and patent DA. Hemodynamically, during the.

intrauterine life, the major part of the oxygen-rich blood coming from the umbilical vein enters the right atrium and passes across the fossa ovalis into the LV, from where it enters the pulmonary artery and eventually the systemic circulation via the open DA. The vascular resistance in the placental system is comparatively lower to that in pulmonary capillaries and thereby facilitates the right-to-left blood flow through the DA into the aorta. After birth, the volume of intercirculatory mixing decides the severity of hypoxemia, which is optimized by balancing the effective pulmonary and systemic blood flows [7, 8]. The effective systemic blood flow is defined as the volume of oxygenated pulmonary venous return reaching the systemic capillary bed, and the effective pulmonary blood flow, the systemic venous return entering the pulmonary capillary system via the intracardiac and extracardiac shunts. To be most efficient, the shunting of blood must be bidirectional, occurring during both systole and diastole. Being a lower pressure system, this happens better at the atrial level. The interventricular and extracardiac shunt flow may be unidirectional as they function in a high-pressure system across high-pressure gradient. Keeping the shunt bidirectional and balanced is important, as a preferential shunting to either side will lead to clinical deterioration.

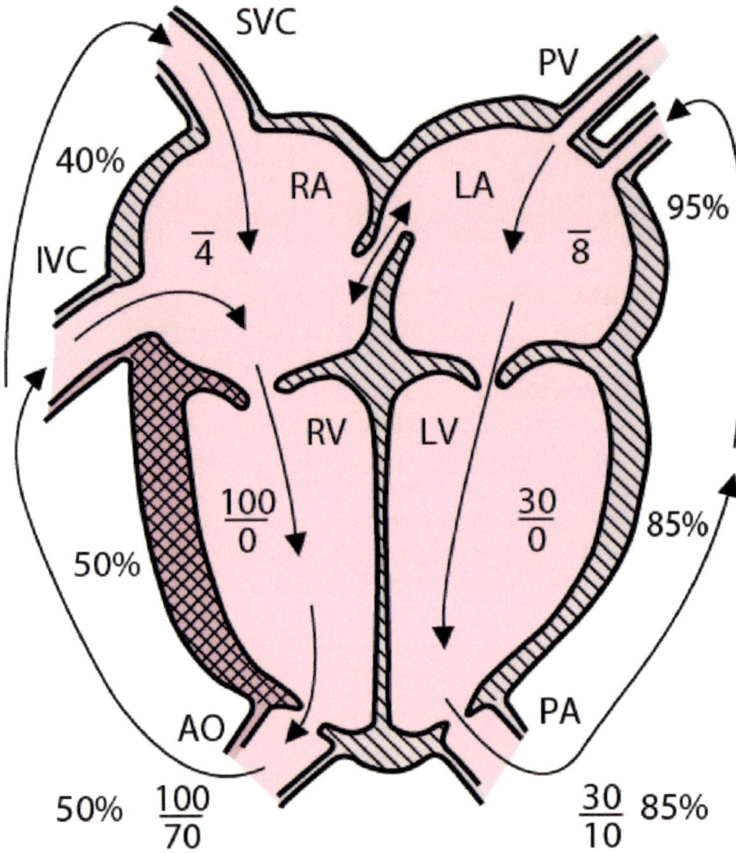

Figure 4.
Hemodynamics in DTGA (curtsey MSD manual).

6. Clinical presentation

Infants with TGA are generally born at term and are normally developed [7, 8, 13]. They may be asymptomatic initially and develop variable cyanosis and respiratory distress shortly after birth. Most patients present with mild shallow tachypnea, which may not be associated with significant retractions. The neonates appear comfortable compared to the degree of cyanosis. Cyanosis is "fixed" and does not respond to oxygen supplementation, resulting in a failed hyperoxia test. If DTGA is uncomplicated, there may not be any murmur. The cases complicated with VSD, PS, and LVOO present with murmurs specific to the lesions. On auscultation, S1 may be loud and single with no audible P2 sound.

The cyanosis in DTGA varies from mild to severe. The degree of mixing between the deoxygenated and oxygenated blood and the presence of other cardiac anomalies determine the severity of cyanosis [7, 8, 13]. If associated with restrictive PFO or ASD, intact ventricular septum or very small VSD severe cyanosis during the newborn period may be noted, which is initially modified by the presence of a patent DA. If a VSD is present, the mixing is better, and the oxygen desaturation not significant; hence the lesion may be detected by a failed pulse oximetry screening at birth. Infants with large VSDs may have no cyanosis or mild cyanosis only when crying or straining. These infants develop CCF over the first month of life. If DA is open and pulmonary vascular pressure low, cyanosis is less intense as the shunting across DA is from left to right. If associated with both VSD and LVOO cyanosis at birth is extreme, the severity depending upon the degree of diminution of pulmonary blood flow due to left ventricular outflow tract obstruction. DTGA associated with VSD and progressively advancing pulmonary vascular obstructive disease is an uncommon variety and presents with progressive cyanosis despite early palliative procedure and absence of CCF. If DTGA is associated with an intact ventricular septum and PDA along with one of the following: pulmonary hypertension, coarctation of the aorta, or interrupted aortic arch, the infant may have higher postductal oxygen saturations than the preductal, as the better-saturated blood flows into the descending aorta via right-to-left ductal shunting. This sometimes presents as differential cyanosis.

About 80% of infants born with TGA are asymptomatic for the first 24–48 hours of life since the DA is open [13]. After discharge, as the DA closes, such babies return with acute cardiorespiratory decompensation with hypoxemia and severe metabolic acidosis due to tissue hypoperfusion. The universal application of the pre-discharge Congenital Cyanotic heart disease pulse oximetry screening test, recommended by the AAP has significantly improved the chances of a timely diagnosis of TGA in neonates, and such situations are now minimized. In some cases, the presentation may be delayed by days or even weeks, but most are revealed within the neonatal period. If untreated, about 30 percent of TGA cases will die in the first week, 50 percent in the first month, and almost 90 percent within the first year of life.

7. Diagnosis

Antenatal diagnosis of TGA by fetal echocardiography has improved since the visualization of outflow tracts to evaluate the major arteries' relationship is regularly performed. Postnatally, any baby failing the CCHD test or presenting with tachypnea and cyanosis that does not improve with 100% oxygen supplementation should raise the suspicion of a shunting cardiac lesion. A Hyperoxia test confirms the fixed shunting and differentiates between cardiac and pulmonary etiology of hypoxia [13].

Echocardiography with Doppler Study is confirmatory. The procedure should include evaluating atrioventricular and ventricular arterial connections and the presence of other anatomical cardiac anomalies, including those of coronary arteries. In D-TGA, two-dimensional echocardiography in the subcostal view demonstrates a great artery arising from the posteriorly situated left ventricle and bifurcating into left and right pulmonary arteries, while in the short-axis or parasagittal view, the aorta is visualized coming out anteriorly from the right ventricle [7, 8, 13, 14]. The atrial evaluation is important to confirm the presence of FO or ASD and the degree of interatrial flow, which determines the critical intercirculatory mixing. The presence of a VSD assures mixing but suggests the possibility of aortic arch anomalies. Complex TGA with VSD and coarctation have worse surgical outcomes and mortality. The presence of patent DA and the degree and direction of shunting blood are important information to obtain. Coronary artery anatomy and its variations must be ascertained via echocardiography, angiography or cardiac magnetic resonance imaging (MRI) prior to surgery. Atrioventricular valvular anomalies and chordae tendineae's relationship with interventricular septum and ventricles are other important features to evaluate before surgery.

Electrocardiography in TGA is generally normal with right-axis deviation and right ventricular hypertrophy. Biventricular hypertrophy may be noted if DTGA is complicated with large VSD, PDA, PS, or LVOO. Chest radiography may show normal or slightly increased heart size or the typical "egg on a string" appearance with

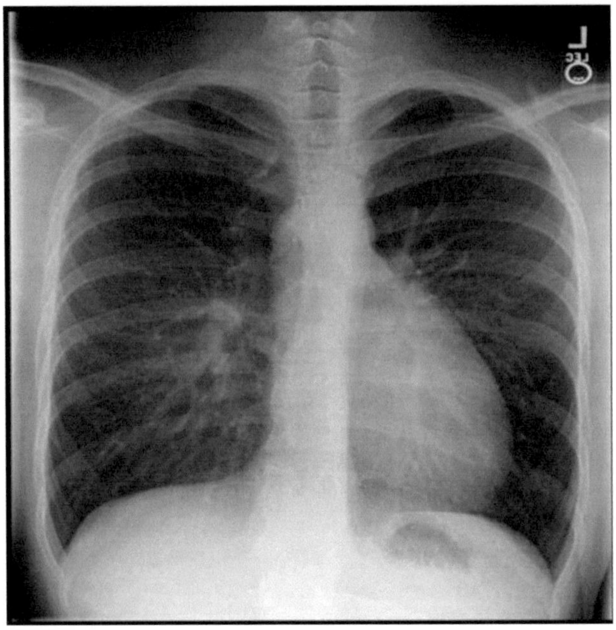

Figure 5.
Chest radiograph in DTHA showing "egg on string appearance".

a narrow mediastinum, created by the anteroposterior positioning of the aorta and pulmonary artery and by involuted thymus (**Figure 5**) [7, 8, 13, 14]. The pulmonary vascular markings may be normal or increased. In the presence of a large VSD or straddling tricuspid valve, the pulmonary flow is increased, and cardiomegaly with congestive heart failure may supervene soon after birth. Cardiac catheterization (CC) and coronary angiography should be done to determine the origin, anatomy, and course of CA, espy if the echocardiography fails to visualize it well before surgery. CC is also used for performing balloon septostomy as a palliative measure to establish mixing of blood.

The conditions which should be clinically differentiated from DTAG are other cyanotic congenital cardiac defects, such as double-outlet right ventricle with malposed great arteries; tricuspid atresia; pulmonary atresia with an intact ventricular septum or with VSD; Tetralogy of Fallot with absent pulmonary valve, pulmonary atresia or severe pulmonary valve stenosis; TAPVR and truncus arteriosus.

8. Management

8.1 Medical management

Once the diagnosis is confirmed after a though evaluation, the focus should be on cardiorespiratory stabilization and adequate systemic oxygenation by ensuring optimum mixing of blood. As the first step, DA's patency is maintained with continuous intravenous PGE1 infusion, which is followed by balloon atrial septostomy (BAS, Rashkind procedure) [15, 16]. The presence of a significantly lower preductal oxygen saturation (SaO2) than the post ductal suggests inadequate atrial shunting and mixing of blood and indicates CC and BAS [17]. BAS is used to increase the atrial level shunt. The procedure can be performed at the bedside under echocardiographic guidance or in the catheterization laboratory under fluoroscopy and echocardiography. It involves accessing the heart via the umbilical or femoral vein and then inserting a deflated balloon across the atrial septum into LA. The balloon is then inflated and pulled back across the septum. The procedure is repeated, and follow-up echocardiography and clinical hemodynamic assessment are done to ensure the formation of an effective ASD and optimum intracardiac mixing.

Intravenous fluid and bicarbonate should be given to correct metabolic acidosis, and pulmonary support should be provided as and if indicated. Congestive heart failure should be treated if present. Appropriate nutrition is an essential part of the management.

8.2 Surgical management and procedures

Surgery is undertaken within the first 1–2 weeks of life. The infants' clinical status and hemodynamics determines the timing. Surgery is generally undertaken after the palliation procedures, but in selected cases, such as DTGA with intact ventricular septum and restrictive foramen ovale with severe metabolic acidosis may be performed without, and within the first few days of life [18]. The arterial switch operation (ASO, Jatene procedure) is the standard surgical procedure for DTGA. ASO has reduced the mortality in DTGA from almost 90% in unoperated infants to <5% in those who undergo surgery [19–21]. The perioperative mortality with ASO in simple DTGA is <1%, while it is 4% in those with complex DTGA. Reportedly a delay of ASO beyond three days after birth may be associated with increased morbidity and health care costs. Other surgical options in DTDA are the Mustard, Senning, and Rastelli procedures. The surgical procedures are selected according to the complexity of DTGA as follows:

a. Simple DTGA with intact VS and no other cardiac anomalies: ASO is the preferred surgery and is performed within the first month of life for the LV to be able to sustain the systemic pressure circulation. In this procedure, the two great arteries are transected and translocated to the opposite root, thus reestablishing the ventriculoarterial concordance. In addition, the coronary arteries are mobilized and reimplanted into the neo-aortic roots with the formation of buttons around them. If coronary arterial reimplantation or mobilization is not possible, the atrial switch operation, known as the Mustard or Senning procedures, may be undertaken. In this operation, the oxygen-rich blood is redirected to RV and Ao, and the oxygen-poor blood to the LV and PA via a two-way interatrial baffle, created by the patient's own tissue or synthetic material. The atrial switch is less preferred due to its late complications. It may be used for palliation in selected complex cases of LTGA.

b. DTGA with VSD: ASO and VSD closure are performed.

c. D-TGA with VSD and LVOO due to pulmonary stenosis (PS): The Rastelli procedure is the standard surgery, and ASO can be done with or without relieving the LVOO obstruction. The decision about choosing one procedure over the other depends on the size of VSD, the severity, and type of LVOO, the status of neo aortic, i.e., pulmonary valve, and the anatomy of coronary arteries. The objective and considerations are to minimize the recurrence risk of LVOO (i.e., PS) or future repeat surgeries and to optimize the pulmonary valve functions. Both ASO and Rastelli procedures involve an arterial switch following which the LV functions as the systemic ventricle and the RV as the pulmonary ventricle [14, 17]. The principle is to redirect the ventricular outflows. Rastelli procedure is undertaken in patients of D-TGA with a large VSD and significant

LVOO. In this procedure, the LV outflow tract obstruction is baffled through the VSD, thus closing it, and the oxygenated blood from the left ventricle is directed into the aorta. A valved conduit is placed between RV and PA, and the deoxygenated blood from the right ventricle enters the PA via the conduit. The Rastelli procedure with a perioperative mortality of <5 percent generally provides better and more durable relief of LVOT obstruction than ASO but has more postoperative complications.

d. TGA with VSD and pulmonary arterial vascular disease: This is a rare type in which surgery may not be beneficial as PA hypertension is progressive. Palliation may be an option in some cases.

8.3 Complications of surgery

Complications may occur in 5 to 25 percent of patients who undergo ASO; the commonest one that will need reintervention is pulmonary artery stenosis [22–26]. If the right ventricular pressure becomes close to systemic levels or if the lung perfusion scan is abnormal catheter-based dilation and stent placement may be performed. Other less common complications include coronary artery insufficiency, neo-aortic root dilation, and neo-aortic regurgitation. With the Rastelli procedure, complications such as conduit stenosis needing replacement, atrial and ventricular arrhythmias, and right and left ventricular failure are reported. Complications of the atrial switch include right ventricular failure, arrhythmias, and baffle-related sequelae [20, 23, 24].

9. Prognosis of TGA

With the introduction of palliation followed by cardiac surgical procedures, the outcome of TGA has changed from a universally fatal disease to a long-term survival rate of over 90% [20]. Patients who undergo ASO have the best long-term survival and the lowest morbidity, and the best functional outcome compared to other surgical procedures. ASO is reported to have a > 95 percent survival at 15 to 25 years post-discharge, whereas survival in the Mustard atrial switch procedure is approximately 80 percent at 20 years and 75 percent at 25 years. [25] In the Rastelli procedure, 10-year survival rates are reported to be 80 to 94 percent [22]. Perioperative mortality is greater in patients with complex DTGA compared to those with simple DTGA. Progressive congestive heart failure and sudden death are the commonest causes of death. The clinical risk factors for mortality in TGA are prematurity, low birth weight, single coronary artery, aortic arch obstruction, and RV hypoplasia. Ninety-three percent of the ASO recipients are reported to be free from long-term cardiovascular complications, such as arrhythmia, cerebrovascular accident, heart failure-related hospitalization. Their cardiorespiratory exercise capability is at 91 percent of normal, while those who undergo atrial switch procedures 75 percent [26]. VSD repair, residual right-sided obstructive lesions, and decreased left ventricular function are the risk factors for poor exercise performance. Those who undergo Mustard or Senning operations are more likely to suffer from arrhythmia and right heart failure.

Neurodevelopmental impairment is common in babies after ASO [27–33]. Surgery at postnatal age > 2 weeks and the presence of VSD are documented to be associated with reduced perioperative brain growth and poor language performance scores at 18 months of age, respectively [15]. In one study, at 16 years of age, 65 percent of the ASO recipients required special education, occupational therapy, psychotherapy, and counseling; and about 20 percent suffered from attention deficit hyperactivity disorder and poor global psychosocial function scores [28]. The occurrence of seizures in the postoperative period is independently associated with lower neuropsychological scores. In another prospective study performed on 45 children with DTGA, the neurodevelopmental testing at age four to six years revealed the scores to be normal for cognition, language, and verbal working memory but lower for inhibition control, cognitive flexibility, social cognition, and executive function, when compared with controls [29, 30]. Neurodevelopmental outcomes are modified by the timing of diagnosis (prenatal versus postnatal), the severity of hypoxemia in the newborn period, the timing of surgery, associated cardiac defects, perioperative complications, necrotizing enterocolitis, and the need for extracorporeal membrane oxygenation. Brain MRI may be abnormal in 20–50% of cases who undergo surgery and demonstrate focal ischemic changes, white matter changes, volume loss, and periventricular leukomalacia [32, 33].

10. Long term outcomes in adult survivors

In the adult survivors of DTGA who undergo atrial switch procedure, RV failure and TR are common as the morphological RV functionally supports the high-pressure systemic circulation instead of the physiological low-pressure pulmonary circulation as it is supposed to [34]. More common long term complications are cardiac arrhythmias, which can result in sudden death due to atrial flutter. In the longer living survivors, sinus node dysfunction with the eventual need for pacemaker placement has been noted. In the adult survivors of ASO recipients, pulmonary arterial stenosis and distortion, neoaortic root dilatation, and aortic regurgitation are the reported complications.

In the patients with LTGA, the congenitally corrected transposition, the systemic high-pressure circulation is supported by the morphological RV and a weak tricuspid valve. In these patients, TR is common which progressively worsens with age, eventually leading to RV failure. Additional surgical interventions may be required if RV dilatation or dysfunction appear. Almost 90% of cases of LTGA are associated with other cardiac defects, such as VSD (70–80%), PA (30–60%), mitral valve anomalies (straddling valve, additional cusps, dysplasia, 55%), and TR (90%). Rhythm disorders, like heart block or atrial arrhythmias, are common in them. Some of these patients may eventually require cardiac transplantation. Appropriate follow-up with periodic hemodynamic and functional assessment is an integral part of these patients' subsequent specialized management for best outcomes [35].

11. Conclusion

TGA is a complex congenital cyanotic cardiac anomaly characterized by ventriculoarterial or ventriculoatrial discordance. DTGA is the commonest cyanotic heart disease occurring during the first week of life. Advancements in early diagnostic measures, medical management, and surgical procedures have increased this former universally fatal disease's survival rate to over 90%. The post-surgical course is complicated by cardiac anatomic, hemodynamic, and rhythm disorders, which may require meticulous follow-up and additional surgical interventions, including cardiac transplants in some cases with LGTA. Long term outcome may be associated with normal to mild neurodevelopmental disabilities depending upon the type, complexity, and course of the disease. Expert follow-up of the patients into adulthood is essential for the best outcomes.

Author details

Rita Prasad Verma
Nassau University Medical Center, East Meadow, NY, USA

*Address all correspondence to: rverma2@numc.edu

References

[1] Mai CT, Isenburg JL, Canfield MA, et al. for the National Birth Defects Prevention Network. National population-based estimates for major birth defects, 2010-2014. Birth Defects Res 2019; 1– 16. https://doi.org/10.1002/bdr2.1589.

[2] Centers for Disease Control and Prevention (CDC). Improved national prevalence estimates for 18 selected major birth defects--United States, 1999-2001. MMWR Morb Mortal Wkly Rep 2006; 54:1301.

[3] Reller MD, Strickland MJ, Riehle-Colarusso T, et al. Prevalence of congenital heart defects in metropolitan Atlanta, 1998-2005. J Pediatr 2008; 153:807.

[4] Tennstedt C, Chaoui R, Körner H, Dietel M. Spectrum of congenital heart defects and extracardiac malformations associated with chromosomal abnormalities: results of a seven year necropsy study. Heart 1999; 82:34.

[5] Momma K. Cardiovascular anomalies associated with chromosome 22q11.2 deletion syndrome. Am J Cardiol 2010; 105:1617.

[6] Becker TA, Van Amber R, Moller JH, Pierpont ME. Occurrence of cardiac malformations in relatives of children with transposition of the great arteries. Am J Med Genet 1996; 66:28.

[7] Fulton DR, Fyler DC. D-Transposition of the Great Arteries. In: Nadas' Pediatric Cardiology, 2nd ed, Keane JF, Lock JE, Fyler DC (Eds), Saunders Elsevier, Philadelphia, PA 2006. p.645.

[8] Wernovsky G. Transposition of the Great Arteries. In: Moss and Adams' Heart Disease in Infants, Children, and Adolescents: Including the Fetus and Young Adult, 7th ed, Allen HD, Shaddy RE, Driscoll DJ, Feltes TF (Eds), Wolters Kluwer Health/Lipincott Williams & Wilkins, Philadelphia, PA 2008. p.1039.

[9] Moene RJ, Oppenheimer-Dekker A, Bartelings MM. Anatomic obstruction of the right ventricular outflow tract in transposition of the great arteries. Am J Cardiol 1983; 51:1701.

[10] Kirklin JW, Barratt-Boyes BG. Complete Transposition of the Great Arteries. In: Cardiac Surgery, Kirklin JW, Barratt-Boyes BG (Eds), Churchill Livingston, New York 1993. p.1383.

[11] Deal BJ, Chin AJ, Sanders SP, et al. Subxiphoid two-dimensional echocardiographic identification of tricuspid valve abnormalities in transposition of the great arteries with ventricular septal defect. Am J Cardiol 1985; 55:1146.

[12] Wernovsky G, Sanders SP. Coronary artery anatomy and transposition of the great arteries. Coron Artery Dis 1993; 4:148.

[13] Rao PS. Diagnosis and management of cyanotic congenital heart disease: part I. Indian J Pediatr. 2009 Jan. 76 (1):57-70.

[14] Rao PS. Management of Congenital Heart Disease: State of the Art—Part II—Cyanotic Heart Defects. Children (Basel). 2019 Apr; 6(4): 54. doi: 10.3390/children6040054

[15] Beitzke A, Suppan CH. Use of prostaglandin E2 in management of transposition of great arteries before balloon atrial septostomy. Br Heart J. 1983;49(4):341-344. doi:10.1136/hrt.49.4.341

[16] Hiremath G, Natarajan G, Math D, Aggarwal S. Impact of balloon atrial

septostomy in neonates with transposition of great arteries. J Perinatol. 2011 Jul;31(7):494-499.

[17] Rao PS. Transposition of the Great Arteries in the Neonate. Neonatology today Newsletter. 2010 August. Volume 5. Issue 8

[18] Nevvazhay T, Chernogrivov A, Biryukov E, Biktasheva L, Karchevskaya K, Sulejmanov S, Kalinicheva J, Artemiev N. Arterial switch in the first hours of life: no need for Rashkind septostomy? Eur J Cardioth Surg. 2012 Sep;42(3):520-523.

[19] Lim JM, Porayette P, Marini D, et al. Associations Between Age at Arterial Switch Operation, Brain Growth, and Development in Infants With Transposition of the Great Arteries. Circulation 2019; 139:2728.

[20] Hutter PA, Kreb DL, Mantel SF, et al. Twenty-five years' experience with the arterial switch operation. J Thorac Cardiovasc Surg 2002; 124:790.

[21] Anderson BR, Ciarleglio AJ, Hayes DA, et al. Earlier arterial switch operation improves outcomes and reduces costs for neonates with transposition of the great arteries. J Am Coll Cardiol 2014; 63:481.

[22] Hazekamp MG, Gomez AA, Koolbergen DR, et al. Surgery for transposition of the great arteries, ventricular septal defect and left ventricular outflow tract obstruction: European Congenital Heart Surgeons Association multicentre study. Eur J Cardiothorac Surg 2010; 38:699.

[23] Jatene AD, Fontes VF, Paulista PP, et al. Anatomic correction of transposition of the great vessels. J Thorac Cardiovasc Surg 1976; 72:364.

[24] Prêtre R, Tamisier D, Bonhoeffer P, et al. Results of the arterial switch operation in neonates with transposed great arteries. Lancet 2001; 357:1826.

[25] Khairy P, Clair M, Fernandes SM, et al. Cardiovascular outcomes after the arterial switch operation for D-transposition of the great arteries. Circulation 2013; 127:331.

[26] Kuebler JD, Chen MH, Alexander ME, Rhodes J. Exercise Performance in Patients with D-Loop Transposition of the Great Arteries After Arterial Switch Operation: Long-Term Outcomes and Longitudinal Assessment. Pediatr Cardiol 2016; 37:283.

[27] Karl TR, Hall S, Ford G, et al. Arterial switch with full-flow cardiopulmonary bypass and limited circulatory arrest: neurodevelopmental outcome. J Thorac Cardiovasc Surg 2004; 127:213.

[28] Hövels-Gürich HH, Seghaye MC, Schnitker R, et al. Long-term neurodevelopmental outcomes in school-aged children after neonatal arterial switch operation. J Thorac Cardiovasc Surg 2002; 124:448.

[29] Culbert EL , Ashburn DA, Cullen-Dean G, Joseph JA, Williams WG, Blackstone EH, McCrindle BW, Congenital Heart Surgeons Society. Quality of life of children after repair of transposition of the great arteries. Circulation. 2003 Aug 19;108 (7):857-862.

[30] Neufeld RE, Clark BG, Robertson CM, et al. Five-year neurocognitive and health outcomes after the neonatal arterial switch operation. J Thorac Cardiovasc Surg 2008; 136:1413.

[31] McGrath E, Wypij D, Rappaport LA, et al. Prediction of IQ and achievement at age 8 years from neurodevelopmental status at age 1 year in children with D-transposition of the great arteries. Pediatrics 2004; 114:e572.

[32] Bellinger DC, Wypij D, Rivkin MJ, et al. Adolescents with d-transposition

of the great arteries corrected with the arterial switch procedure: neuropsychological assessment and structural brain imaging. Circulation 2011; 124:1361.

[33] DeMaso DR, Labella M, Taylor GA, et al. Psychiatric disorders and function in adolescents with d-transposition of the great arteries. J Pediatr 2014; 165:760.

[34] Breinholt JP , Sheba John Management of the Adult with Arterial Switch. Methodist Debakey Cardiovasc J. Apr-Jun 2019;15(2):133-137. doi: 10.14797/mdcj-15-2-133.

[35] Safi LM, Bhatt AB. Update on the Management of Adults With Arterial Switch Procedure for Transposition of the Great Arteries. Curr Treat Options Cardiovasc Med. 2017 Jan;19(1):4. doi: 10.1007/s11936-017-0505

Chapter 9

Genetics of Sirenomelia, the Mermaid Syndrome

Rita Prasad Verma

Abstract

Sirenomelia (SML) is a rare, almost universally fatal congenital malformation presenting pathognomically with fused lower extremities and absent or malformed perineum. The classic Sirenomelia sequence includes a uniform spectrum of caudal malformations, spinal defects, and a single umbilical artery. SML is postulated to be due to a genetic predisposition, unmasked by biochemical or environmental triggers. Primary developmental defects in the formation of caudal mesoderm or embryonic caudal vessels with resultant local tissue hypoperfusion are proposed hypotheses for its pathogenesis. SML occurs sporadically in humans, presumably due to a spontaneous mutation, and is speculated to have an autosomal dominant inheritance pattern. In mutant mice, specific defects in Cyp26a1 and Bmp 7 genes are demonstrated to produce offsprings with SML. Bmp 7 is a signaling protein, which belongs to the transforming growth factor-β (TGF β) superfamily. Tsg 1, a Bmp and chordin-binding protein, functions as an activator-inhibitor of Bmp signaling in the embryonic caudal region (ECR). Loss of Bmp7 genes combined with a complete loss or half-dose of Tsg 1 is demonstrated to produce an invariable SML phenotype. SML is also demonstrated to occur with increased Retinoic acid (RA) signaling in the ECR. The Cyp26a1 gene is involved in coding for an enzyme, which expresses in ECR and degrades RA. A specific defect in this gene leads to excess local RA concentration and SML generation with a reported 20% penetrance in mutant mice. However, the mutational screening of Cyp26a1 and Bmp 7genes has failed to confirm their involvement in mankind and the molecular defect and genetic inheritability of SML in humans remain undefined.

Keywords: Sirenomelia, Blastogenesis, Vascular steal hypothesis, Retinoic acid, BMP 7 signaling, Cyp26a1gene

1. Introduction

Sirenomelia (SML) is a rare and almost universally fatal congenital malformation characterized by the pathognomonic feature of fused lower extremities, with absent or malformed perineum and Potter facies (**Figure 1**) [1]. SML, also called mermaid syndrome, is one of the most striking phenotypes among human anomalies. The exact etiopathogenesis of SML is undetermined, and the syndrome is postulated to be due to a genetic predisposition that is unmasked by a biochemical or environmental trigger factor. Several hypotheses have been proposed for the pathogenesis, among which the most accepted ones are vascular steal phenomenon, defective blastogenesis, and mechanical compression of the fetal caudal body. [1–8] Studies in mutant mice have provided significant and relevant information towards the understanding

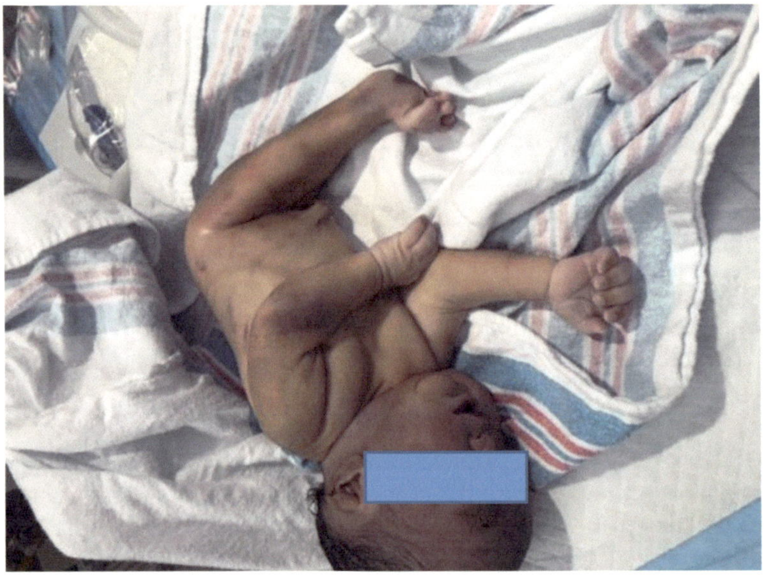

Figure 1.
Sirenomelia in a newborn infant showing merged hind extremities and Potter facies.

of the genetic aspect of the anomaly. [1] SML occurs sporadically, and even though not recognized as familial, it is documented to be more common in twin monozygotic pregnancies, among whom the relative risk is stated to be increased by 100-fold. About 9–15% of all cases of SML are products of twin gestation. Environmental and teratogenic factors, such as cocaine, retinoic acid, heavy metals, cyclophosphamide, and certain antibiotics, have been linked to SML in humans and animal models [1, 9]. In addition, nicotine, alcohol, radionuclides, diethylpropion- an appetite suppressor, organic solvents of fats, and even air pollution have been associated with SML and caudal regression syndrome, which is controversially considered as its minor form [4, 9]. Other authors have reported fetal exposure to cadmium, lithium, phenytoin, sodium valproate, carbamazepine, warfarin, methylergonovine, diethylpropion, trimethoprim, and ochratoxin-a type of fungus as possible triggers for the anomaly. [7–9] Some of the maternal complications, such as diabetes mellitus, hyperthermia during the 1st trimester of pregnancy, amniotic bands, and age below 20 years or over 40 years at conception, have also been implicated in the pathogenesis of SML. [1, 9]. Recently a case report tentatively associated persistent early gestational maternal Chlamydia trachomatis (CT) infection with SML. CT is an obligate intracellular pathogen that is recognized for its cytopathogenic effects like cellular disruption, tissue dysgenesis, and genomic instability and is known to invade the placenta and cause fetal demise. [10] SML has no ethnic or geographical preferences, and a gender preponderance with a male to female ratio of 2.7:1 has been described. [1] The prevalence of SML is reported to be between 1.1 and 4.2 per 100,000 births [1, 6]. The molecular basis of the defects in SML is yet to be defined.

2. The phenotype

The typical phenotype of SML is characterized by a partial or complete merging of the lower extremities into one single limb. The deformity, which gives the mermaid appearance to the syndrome, is invariably associated with smooth perineum, malformed or absent perineal structures, and in most instances, Potter facies (**Figure 1**). The classic Sirenomelia sequence consists of a uniform spectrum of visceral malformations involving urogenital and gastrointestinal systems, pelvic and lumbosacral spinal defects, and a single aberrant umbilical artery. Renal agenesis or severe renal malformation are almost invariably present and lead to oligohydramnios and consequent pulmonary hypoplasia. Other commonly observed abnormalities in SML are sacral agenesis, abnormal vertebrae, hemivertebrae, meningomyelocele, and less frequently, cleft palate, cardiac defects omphalocele, and pentalogy of Cantrell [1, 11]. It is hypothesized that a defect in blastogenesis leads to impaired angiogenesis in the embryonic caudal region, and the resultant vascular compromise causes anomalous development of the area structures. The lower extremity phenotype is evidenced to occur due to merger and not by fusion of the two bilaterally positioned fetal hind limb buds as suggested by the histopathological persistence of epithelial linings in the conjoined limb fields. [1] The merged lower extremities are abnormally rotated by 180 degrees to the normal position of legs (**Figure 1**). Normally, during development, the hind limb buds rotate medially, and the original ventral surface turns to become dorsal [12] In SML, due to the early midline posterior fusion of the embryonic hind limbs, this rotation does not occur, as a result of which, the soles of the affected fetus face anteriorly, and the fibulae end up being placed medially between the tibiae. [1] Some researchers believe that the typical phenotype in SML is secondary and not a primary defect. As the developing leg fields in the fetus normally lie along the lateral body wall, it is thought that a developmental failure of caudal midline structures, such as ventrally positioned cloaca and urogenital sinus, and dorsally positioned somites and neural tube, results in the approximation and final merging of the two analgen of fetal limb buds [12, 13]. In SML, kidneys are almost always affected, and although the gonads may be spared, the urethra is undeveloped and external genitalia either absent or represented by abnormal tissue tags [14]. Other almost invariably present anomalies involve the gastrointestinal tract and are represented by a blind-ending colon, rectal atresia, and imperforate anus [15]. In the classic form, SML is fatal during the perinatal period, whereas, in the minor forms, reconstructive pelvic and limb surgery may prolong the life if the renal function is compatible with survival. [16] Notably, the neurodevelopment of the surviving babies is reported to be normal.

3. Pathogenesis

Although unsubstantiated, it is believed that SML is an autosomal-dominant genetic condition in humans, and each case is the result of a new spontaneous mutation. It is speculated that the syndrome is induced by a combination of genetic and environmental influences. The most accepted non-genetic hypotheses for the pathogenesis of SML are the vascular steal phenomenon and defective blastogenesis. These two theories are not mutually exclusive as any deficiency in blastogenesis could potentially lead to fetal vascular and visceral maldevelopment. Genetically, SML phenotype has been demonstrated in mutant mice with both gain-of-function of retinoic acid (RA) signaling and loss-of-function of bone morphogenetic protein (Bmp) signaling [11].

3.1 Non-genetic theories

3.1.1 Vascular steal hypothesis

SML is characterized by an anomalous single umbilical artery that diverts blood away from the caudal fetal limb buds to the placenta, thereby compromising the regional blood supply during the early developmental period. In the early fetal stage, the vitelline artery complex serves as an embryonic vascular network that supplies the yolk sac. [2, 11] In SML, the umbilical vasculature develops abnormally and connects to the vitellus in an anomalous way, resulting in the formation of a large aberrant aorta-like vessel that originates from the vitelline artery high in the future abdominal cavity. The normal aortic branches, such as the celiac artery, are either separated from this abnormal abdominal aorta or are absent or hypoplastic. This large anomalous vessel functions as a single umbilical artery and diverts blood away from the lower part of the fetus to the placenta, thus severely limiting the blood and nutrients supply to the mesoderm of the caudal fetal body. In autopsies, the tissues dependent upon the lower branches of this aberrant steal aorta are found to be malformed or in different stages of incomplete or arrested development. However, case reports of SML with two umbilical arteries have been described, and an aberrant single umbilical artery, which was considered as the hallmark of SML and a major differentiating feature from caudal dysgenesis, is now not considered to be so [2, 15, 17–19].

3.1.2 Defective blastogenesis hypothesis

According to this theory, SML results due to an impairment in blastogenesis during the final stages of gastrulation. A primary defect in blastogenesis leads to abnormal angiogenesis and poor blood supply to the embryonic caudal region at the tailbud stage during the third human gestational week with consequent deficiencies in the area structures' growth and development (**Figure 2**) [18]. Presumably, the timing, duration, and severity of the disruption determine the clinical presentation's phenotypical variability. Developmentally, during the gastrulation phase, epiblast cells move massively through the primitive streak and transform a two-layered blastocyst into a three germ layered embryo, with the endomesoderm emerging in a rostrocaudal sequence [20]. At late gastrulation, the regressing primitive streak forms the tailbud, which is composed of a mass of loose mesenchyme covered by ectoderm [13, 21]. As the tailbud grows, the remnants of the primitive streak continue to involute ventrally, eventually forming a distal thickened ectodermal area called the ventral ectodermal ridge (VER) [21], which basically is a continuation of the posterior primitive streak.

Meanwhile, the surface mesoderm precursors continue to internalize and contribute to the tail elongation and caudal structures formation. This process is confirmed in both chick and mouse embryos by performing analysis of the concerned cells' molecular markers [22]. As demonstrated in chicks, the VER cells can undergo an epithelial-mesenchymal transition process and thus accumulate mesoderm cells in the lateral and ventral tailbud region. The VER controls cell proliferation in the underlying mesoderm and is presumably the signaling center responsible for tail elongation [22]. A disruption in this embryological process is postulated to produce the SML phenotype. The questions arise about the triggering factors' identity and source in the induction of the specific blastogenesis defect. These factors could be environmental or more recently recognized pathological condition of excess accumulation of retinoic acid and its metabolites in the embryonic caudal buds, which hypothetically may superimpose upon an unidentified genetic susceptibility.

3.1.3 Mechanical compression of embryonic caudal region

Abnormalities in the amniotic forces may adversely impact the embryo's caudal body and lead to its hypoplasia. It is also suggested that abnormal excessive distension of caudal neural tube may lead to lateral rotation of the developing mesoderm and cause fusion of lower limbs and abnormal closure of primitive gut and

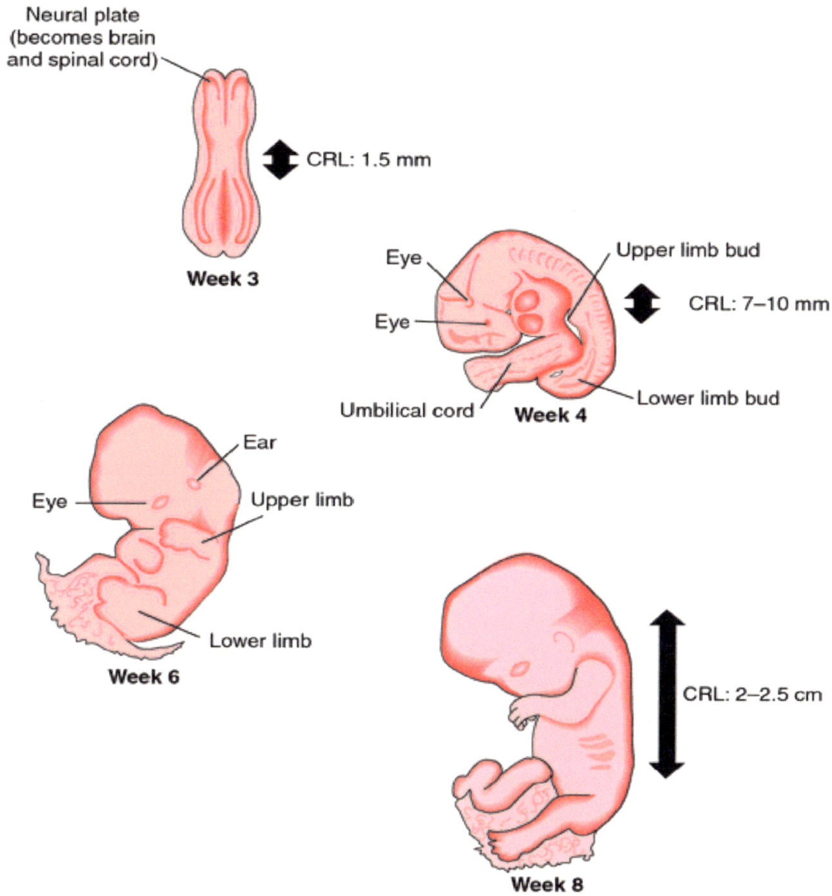

Figure 2.
Diagrammatic representation of the early development during 3–8 weeks after fertilization in humans. CRL = crown to rump length.

urethra. [1] Finally, it is speculated that the merger of the lower extremity in SML is a consequence of the allantois abnormally intervening or inhibiting the cleavage of the lower limb bud field into two lateral masses. These hypotheses, however, are not confirmed.

3.1.4 Environmental factors

RA, maternal diabetes mellitus and heavy metals have been reported to be important environmental risk factors for caudal malformations. In a mouse model, RA exposure resulted in multiple caudal structural anomalies, such as tail agenesis, caudal vertebral defects, spina bifida occulta or aperta, imperforate anus, rectovesical or rectourethral fistula, renal malformations, cryptorchidism, gastroschisis, limb malformations, and the classical SML phenotype. [23] On serial histopathological examinations, progressive hemorrhage, edema, cell death, vascular disruption, and tissue deficiencies were demonstrated as the adverse effects of fetal RA exposure. The embryos treated with RA displayed early cell death, particularly in the caudal median axis, hindgut, and neural tube, and a failure to develop the tailbud. The symmelia appeared to be due to the failure of fission or the merger of limb fields rather than a fusion of two limb buds [1, 23]. As RA levels can be modified by genetic, nutritional, and iatrogenic factors, their involvement in the causation of SML could occur via multiple pathogenic routes. Still, regardless of RA's association in the genesis of caudal anomalies in mice models, RA has not been so far documented to induce or trigger SML in humans. Also, even though maternal diabetes mellitus is associated significantly with caudal dysgenesis, its association with SML is not confirmed. Only 0.5–3.7% of SML cases are reported to have mothers with the disorder as opposed to about 15% in caudal dysgenesis [9]. Exposure to heavy metals is another group of environmental toxins, and their anecdotal and unconfirmed association with SML in humans and experimental animals has been described. [24–26].

3.2 Genetic factors

The genetic defects specific to SML have not been so far identified in humans. Except for a few recently published cases, all human fetuses with SML are reported to have normal karyotypes. Gabriele et al. described a fetus with SML having a triploid mosaic (69, XXX/46, XX), [27] while another case was noted to have a de novo balanced reciprocal translocation 46X, t(X; 16) (p11.23; p12.3), in who the chromosomal breakpoints on the pairs of chromosomes did not disrupt the coding genes associated with early blastogenesis [28]. A big breakthrough in the genetic pathogenesis of SML was brought about by the experimental findings in mutant mice. While no genetic inheritability was established in humans, a genetic component was first detected in the SML phenotypical offsprings of mice carrying mutations at or near the T locus in the brachyury gene (short-tail strain) and in the axin1 gene (fused strain), the two genes that are involved in the structural development of tail and caudal body [1, 11]. Another spontaneous mutation called sirenomelia (*srn*) was identified to cause hind limb fusion in homozygous mice [1, 29]. In humans, mutations in the homeobox-containing gene HLXB9 gene is the only recorded genetic aberration associated with congenital caudal anomalies seen in the autosomal dominant sacral agenesis of Currarino syndrome, which presents with pelvic malformations, anal atresia, meningomyelocele, and urogenital defects but not with SML [30].

In experimental mutant mice mothers, defects in the Cyp26a1 and bone morphogenic protein7 (Bmp 7) genes are identified to induce a phenotypical SML product of conception. [1, 11] The Cyp26a1 gene is involved in coding for the enzyme that breaks down RA. RA effectuates vasculature development in the fetal caudal region, but its excessive accumulation disrupts the process. Bmp7 is an important protein related to angiogenesis and stimulates the production of endothelial cells, vessels, and tissue in the fetal caudal region in order to promote normal growth of lower extremities. Disruption of the Cyp26a1 gene is demonstrated to result in incomplete development of the embryonic caudal region and present with hind limb fusion in mice, which is also seen in mice with knockouts or mutations in both Tsg1 and Bmp7 [11].

3.2.1 Decreased Bmp signaling in the embryonic caudal region

An important link between Bmp7 and twisted gastrulation (Tsg) was reported by Zakin et al., who noted that the loss of Bmp7 combined with a complete loss or half-dose of Tsg produces an invariable SML phenotype in mutant mice models. [31]. Bmp 7 is a member of the secreted multifunctional signaling proteins, which belong to the transforming growth factor-β (TGF β) superfamily. Tsg can function as an activator of the inhibitor of Bmp signaling in the caudal embryonic region. [32] Tsg is a Bmp and chordin-binding protein that has multiple effects on BMP metabolism in the extracellular space. Bmp7 is one of many Bmps that bind to Tsg.

In mice, molecular marker studies indicate that the SML phenotype is associated with a defect in ventroposterior mesoderm formation. In Xenopus, a species of frogs, co-injection of Tsg and Bmp7 morpholino oligonucleotides showed a strong synergistic effect in substantially inhibiting the formation of ventral mesoderm, suggesting that the dorsoventral patterning of the mouse posterior mesoderm is regulated by Bmp signaling. [1, 11] It is experimentally evidenced that the single Bmp7 mutant does not generate a mermaid phenotype and the mermaid phenotype in the Bmp7; Tsg double mutant results from a reduction in Bmp signaling in the ventral caudal mesoderm. In zebrafish, deficient Bmp signaling after the mid-gastrula stage causes deficiencies in ventral mesoderm formation, with defects in kidney and excretory system morphogenesis, e.g., aberrant cloaca [33]. Thus, among the mutant mice models that have been characterized so far, the Cyp26a1 and Bmp7; Tsg engineered mice mutants are considered to be the best models to investigate the pathogenesis of SML in humans as they exhibit a phenotype that is closest to the human presentation.

Bmp signaling has important roles in early embryogenesis. Bmp ligands and their extracellular antagonists and modulators control gastrulation. During late gastrulation, in mice, several Bmp ligands and their extracellular antagonists and modulators are expressed in dynamic and partially overlapping domains [34, 35] such as Bmp2 and Bmp7 in the VER, and Bmp4, Bmp7, and the Bmp antagonist Noggin in the underlying mesoderm. These have experimentally demonstrated roles in the embryonic tailbud's growth by promoting the formation of caudal mesoderm in the VER in mice [36]. Furthermore, the termination of cell movements through the VER coincides with Bmp signaling attenuation in chicks. [36] Besides, the double Bmp7; Tsg mutant SML phenotype strongly supports Bmp signaling involvement in caudal development [31]. Physiologically, Bmp signaling is demonstrated to promote endothelial cell activation, migration, and proliferation, thereby contributing to angiogenesis, vasculogenesis, and normal vasculature remodeling of the primitive capillary plexus [1, 36, 37]. Thus, Bmp signaling is critical for the normal formation of the mesoderm and differentiation of the hematopoietic and endothelial precursor cells.

3.2.2 Increased RA signaling in embryonic caudal region

RA signaling in the genesis of SML is well established in experimental animals. RA, the active metabolite of vitamin A, is degraded by the enzyme Cyp26a1, which is specifically expressed in the embryonic caudal region and the developing vascular network. [38, 39]. A deficiency of Cyp26a1 is demonstrated to result in excessive RA activity in the embryonic caudal region and induce multiple caudal defects, including that of SML with a reported 20% penetrance. This enzyme's lack or deficiency also results in diminished bone morphogenetic protein signaling in the caudal region of the embryo. Cdx2 is a transcription factor that encodes and activates the Cyp26a1 promoter, and Por encodes an enzyme that is required for the function of the Cyp26 family of enzymes. [1, 40, 41]. Even the disruption of these related factors Cdx2, and Por, are noted to result in SML. The relationship is further corroborated by the fact that a Cyp26a1 phenotype can be rescued by decreasing the production of RA by reducing the level of the enzyme haploinsufficiency of Raldh2, which is necessary for its production. [41].

Embryologically, the developing cells in caudal structures are especially sensitive to RA signaling during gastrulation. Its levels are tightly controlled by the expression of its metabolizing enzymes [1] A multitude of experiments that modified the level of RA signaling by genetic and nutritional means have demonstrated that the embryo is particularly sensitive to deviations from normal levels of RA during gastrulation [1, 42–44]. The expression of Cyp26a1 takes place at the early gastrula stage in the primitive streak and the nascent mesoderm; and at the late gastrula stage in the neuropore, hindgut endoderm, and tailbud mesoderm [37, 39, 40]. RA signaling is generated by Raldh2 in the somites in association with specific growth factors. RA plays a key role in the proliferation and differentiation of precursor cells, and Cyp26a1 expression at the caudal area is pivotal in maintaining the physiologically appropriate RA levels. [45, 46] RA excess is shown to negatively regulate endothelial cell proliferation and impede vascular remodeling by inducing premature coalescence and differentiation of precursor cells. In animal models, abnormal development of umbilical and vitelline arteries in embryos similar to those in human SML was noted when pregnant rats were given RA. [47] However, despite strong evidence of Cyp26a1 being instrumental in the generation of SML phenotype, a mutational screening of the Cyp26a1 gene has not confirmed its involvement in caudal dysplasia in humans [48].

Even though both Bmp and RA signaling pathways are critical participants in the development of caudal structural during the early embryonic stage, it is unclear if their effects are synergistic or antagonistic and whether they modulate each other's roles. RA had been shown to decrease Bmp signal duration by reducing the level of phosphorylated Smad1, an intracellular component of the Bmp signaling pathway in the developing neural tube [1, 49], whereas Bmp signaling has been demonstrated to adversely regulates RA signaling during chondrogenesis [50]. However, such effects have not been studied or confirmed to be operative in the generation of SML phenotype. The detection of possible crosstalk between the two signaling cascades might provide helpful information regarding the pathogenesis of SML phenotype in humans.

4. Conclusion

Sirenomelia is a major, almost universally fatal human anomaly with obscure etiopathogenesis. An indeterminate genetic predisposition in combination with exposure to potentially adverse environmental trigger factors is thought to be instrumental in producing the phenotype. Several causal hypotheses have been forwarded, and studies in mutant mice have provided important insight into its generation at embryonic cellular and biochemical levels. Genetic aberrations in the Bmp and RA signaling pathways have been demonstrated to induce SML-like phenotype in the mutant mice. Defining the precise genetics, the roles of combined Bmp and RA signaling pathways, and the unidentified molecular defect in SML are subjects of future research on the subject.

Author details

Rita Prasad Verma
Nassau University Medical Center, East Meadow, NY, USA

*Address all correspondence to: rverma2@numc.edu

References

[1] Garrido-Allepuz C, Haro E, González-Lamuño D, Martínez-Frías ML, Bertocchini F, Ros MA. A clinical and experimental overview of sirenomelia: insight into the mechanisms of congenital limb malformations. Dis Model Mech. 2011 May;4(3):289-99. doi: 10.1242/dmm.007732.

[2] Stevenson RE, Jones KL, Phelan MC, Jones MC, Barr M Jr, Clericuzio C, Harley RA, Benirschke K. Vascular steal: the pathogenetic mechanism producing sirenomelia and associated defects of the viscera and soft tissues. Pediatrics. 1986 Sep;78(3):451-457.

[3] Stocker JT, Heifetz SA. Sirenomelia. A morphological study of 33 cases and review of the literature. Perspect Pediatr Pathol. 1987;10:7-50.

[4] Samal SK, Rathod S. Sirenomelia: The mermaid syndrome: Report of two cases. J Nat Sci Biol Med. 2015 Jan-Jun;6(1):264-266. doi: 10.4103/0976-9668.149227.

[5] Lucas L. Boer Eva Morava Willemijn M. Klein Annelieke N. Schepens-Franke Roelof Jan Oostra Sirenomelia: A Multi-systemic Polytopic Field Defect with Ongoing Controversies. special birth defects. 2017 Volume 109, Issue 10 Pages: 719-804

[6] Fadhlaoui A, Khrouf M, Gaigi S, Zhioua F, Chaker A. The sirenomelia sequence: a case history. Clin Med Insights Case Rep. 2010;3:41-49. doi:10.4137/ccrep.s5347

[7] Valenzano M, Paoletti R, Rossi A, et al. Sirenomelia. Pathological features, antenatal ultrasonographic clues and a review of current embryogenic theories. Hum Reprod Update. 1999;5(1):82-86.

[8] Stockers JT, Heifetz SA. Sirenomelia : a morphological study of 33 cases and review of the literature. Perspect Pediatr Pathol. 1987;10:7-50.

[9] Taee N, Tarhani F, Goodarzi MF, Safdari M, Bajelan A. Mermaid Syndrome: A Case Report of a Rare Congenital Anomaly in Full-Term Neonate with Thumb Deformity. AJP Rep. 2018;8(4):e328-e331. doi:10.1055/s-0038-1669943

[10] Fuchs G, Dianova E, Patel S, Kamanda S, Verma RP. Sirenomelia and maternal chlamydia trachomatis infection: a case report and review. Fetal Pediatr Pathol. 2019 Dec;38(6):524-533. doi: 10.1080/15513815.2019.1627623.

[11] Garrido-Allepuz C, González-Lamuño D, Ros MA (2012) *Sirenomelia* Phenotype in *Bmp7;Shh* Compound Mutants: A Novel Experimental Model for Studies of Caudal Body Malformations. PLOS ONE 7(9): e44962. https://doi.org/10.1371/journal.pone.0044962

[12] Schoenwolf, G., Bleyl, S., Brauer, P. and Francis-West, P. (2009). Development of the limbs. In *Larsen's Human Embryology* (ed. Schoenwolf, G., Bleyl, S., Brauer, P. and Francis-West, P.), pp. 617-644. Philadelphia, Pennsylvania: Churchill Livingstone/Elsevier.

[13] O'Rahilly R, Müller F. Interpretation of some median anomalies as illustrated by cyclopia and symmelia. Teratology. 1989 Nov;40(5):409-421. doi: 10.1002/tera.1420400502.

[14] Källén B, Winberg J. Caudal mesoderm pattern of anomalies: from renal agenesis to sirenomelia. Teratology. 1974 Feb;9(1):99-111. doi: 10.1002/tera.1420090113.

[15] Thottungal, A. D., Charles, A. K., Dickinson, J. E. and Bower, C. (2010). Caudal dysgenesis and sirenomelia-

single centre experience suggests common pathogenic basis. Am. J. Med. Genet. 152A, 2578-2587.

[16] Messineo A, Innocenti M, Gelli R, Pancani S, Lo Piccolo R, Martin A. Multidisciplinary surgical approach to a surviving infant with sirenomelia. Pediatrics. 2006 Jul;118(1):e220-e223. doi: 10.1542/peds.2005-3001. Epub 2006 Jun 26.

[17] Heifetz SA. Single umbilical artery. A statistical analysis of 237 autopsy cases and review of the literature. Perspect Pediatr Pathol. 1984 Winter;8(4):345-378.

[18] Opitz JM, Zanni G, Reynolds JF Jr, Gilbert-Barness E. Defects of blastogenesis. Am J Med Genet. 2002 Dec 30;115(4):269-286. doi: 10.1002/ajmg.10983.

[19] Sara M. Duesterhoeft Linda M. Ernst Joseph R. Siebert Raj P. Kapur Five cases of caudal regression with an aberrant abdominal umbilical artery: Further support for a caudal regression–sirenomelia spectrum First published: 28 November 2007https://doi.org/10.1002/ajmg.a.32028Volume 143A, Issue 24 Pages: 2851-3327

[20] Voiculescu O, Bertocchini F, Wolpert L, Keller RE, Stern CD. The amniote primitive streak is defined by epithelial cell intercalation before gastrulation. Nature. 2007 Oct 25;449(7165):1049-1052. doi: 10.1038/nature06211.

[21] GRUNEBERG H. A ventral ectodermal ridge of the tail in mouse embryos. Nature. 1956 Apr 28;177(4513):787-788. doi: 10.1038/177787b0.

[22] Gofflot F, Hall M, Morriss-Kay GM. Genetic patterning of the developing mouse tail at the time of posterior neuropore closure. Dev Dyn. 1997 Dec;210(4):431-445.

[23] Padmanabhan R. Retinoic acid-induced caudal regression syndrome in the mouse fetus. Reprod Toxicol. 1998 Mar-Apr;12(2):139-151. doi: 10.1016/s0890-6238(97)00153-6.

[24] Orioli IM, Mastroiacovo P, López-Camelo JS, Saldarriaga W, Isaza C, Aiello H, Zarante I, Castilla EE. Clusters of sirenomelia in South America. Birth Defects Res A Clin Mol Teratol. 2009 Feb;85(2):112-118. doi: 10.1002/bdra.20492.

[25] Ferm VH, Carpenter SJ. The relationship of cadmium and zinc in experimental mammalian teratogenesis. Lab Invest. 1968 Apr;18(4):429-32. PMID:.

[26] Hilbelink DR, Kaplan S. Sirenomelia: analysis in the cadmium- and lead-treated golden hamster. Teratog Carcinog Mutagen. 1986;6(5):431-440. doi: 10.1002/tcm.1770060509.

[27] Gabriele T, Gianpaolo G. Sirenomelia: a review on embryogenic environmental theories,novel three-dimensional ultrasound imaging and first trimester diagnosis in a case ofmosaic 69, X .

[28] Kurosawa K, Tanoshima-Takei M, Yamamoto T, Ishikawa H, Masuno M, Tanaka Y, Yamanaka M. Sirenomelia with a de novo balanced translocation 46,X,t(X;16)(p11.23;p12.3). Congenit Anom (Kyoto). 2012 Jun;52(2):106-10. doi: 10.1111/j.1741-4520.2011.00326.x.

[29] Hoornbeek FK. A gene producing symmelia in the mouse. Teratology. 1970 Feb;3(1):7-10. doi: 10.1002/tera.1420030103.

[30] Belloni E, Martucciello G, Verderio D, et al. Involvement of the HLXB9 homeobox gene in Currarino syndrome. Am J Hum Genet. 2000;66(1):312-319. doi:10.1086/302723

[31] Zakin L, Reversade B, Kuroda H, Lyons KM, De Robertis EM. Sirenomelia in Bmp7 and Tsg compound mutant mice: requirement for Bmp signaling in the development of ventral posterior mesoderm. Development. 2005 May;132(10):2489-2499. doi: 10.1242/dev.01822.

[32] Larrain, J., Oelgeschläger, M., Ketpura, N. I., Reversade, B., Zakin, L. andde Robertis, E. M. (2001). Proteolytic cleavage of Chordin as a switch forthe dual activities of Twisted gastrulation in BMP signaling. Development128,4439-4447.

[33] Ujwal J. Pyati, Mark S. Cooper, Alan J. Davidson, Alexei Nechiporuk, David Kimelman Sustained Bmp signaling is essential for cloaca development in zebrafish.Development 2006 133: 2275-2284; doi: 10.1242/dev.02388

[34] Winnier G, Blessing M, Labosky PA, Hogan BL. Bone morphogenetic protein-4 is required for mesoderm formation and patterning in the mouse. Genes Dev. 1995 Sep 1;9(17):2105-2116. doi: 10.1101/gad.9.17.2105.

[35] Fujiwara T, Dunn NR, Hogan BL. Bone morphogenetic protein 4 in the extraembryonic mesoderm is required for allantois development and the localization and survival of primordial germ cells in the mouse. Proc Natl Acad Sci U S A. 2001 Nov 20;98(24):13739-13744. doi: 10.1073/pnas.241508898.

[36] Sho Ohta, Kentaro Suzuki, Katsuro Tachibana, Hideaki Tanaka, Gen Yamada. Cessation of gastrulation is mediated by suppression of epithelial-mesenchymal transition at the ventral ectodermal ridge.Development 2007 134: 4315-4324; doi: 10.1242/dev.008151

[37] Robert J. Garriock, Catherine Czeisler, Yasuo Ishii, Alicia M. Navetta, Takashi Mikawa. An anteroposterior wave of vascular inhibitor downregulation signals aortae fusion along the embryonic midline axis. Development 2010 137: 3697-3706; doi: 10.1242/dev.051664

[38] Abu-Abed S, Dollé P, Metzger D, Beckett B, Chambon P, Petkovich M. The retinoic acid-metabolizing enzyme, CYP26A1, is essential for normal hindbrain patterning, vertebral identity, and development of posterior structures. Genes Dev. 2001 Jan 15;15(2):226-240. doi: 10.1101/gad.855001.

[39] Abu-Abed S, Dollé P, Metzger D, Wood C, MacLean G, Chambon P, Petkovich M. Developing with lethal RA levels: genetic ablation of Rarg can restore the viability of mice lacking Cyp26a1. Development. 2003 Apr;130(7):1449-1459. doi: 10.1242/dev.00357.

[40] Pennimpede T, Cameron DA, MacLean GA, Li H, Abu-Abed S, Petkovich M. The role of CYP26 enzymes in defining appropriate retinoic acid exposure during embryogenesis. Birth Defects Res A Clin Mol Teratol. 2010 Oct;88(10):883-894. doi: 10.1002/bdra.20709.

[41] Niederreither K, Subbarayan V, Dollé P, Chambon P. Embryonic retinoic acid synthesis is essential for early mouse post-implantation development. Nat Genet. 1999 Apr;21(4):444-448. doi: 10.1038/7788.

[42] Duester G. Retinoic acid synthesis and signaling during early organogenesis. Cell. 2008 Sep 19;134(6):921-931. doi: 10.1016/j.cell.2008.09.002.

[43] Niederreither K, McCaffery P, Dräger UC, Chambon P, Dollé P. Restricted expression and retinoic acid-induced downregulation of the retinaldehyde dehydrogenase type 2 (RALDH-2) gene during mouse development. Mech Dev. 1997

Feb;62(1):67-78. doi: 10.1016/s0925-4773(96)00653-3.

[44] Niederreither K, Subbarayan V, Dollé P, Chambon P. Embryonic retinoic acid synthesis is essential for early mouse post-implantation development. Nat Genet. 1999 Apr;21(4):444-448. doi: 10.1038/7788.

[45] Niederreither K, Abu-Abed S, Schuhbaur B, Petkovich M, Chambon P, Dollé P. Genetic evidence that oxidative derivatives of retinoic acid are not involved in retinoid signaling during mouse development. Nat Genet. 2002 May;31(1):84-88. doi: 10.1038/ng876.

[46] Young T, Rowland JE, van de Ven C, Bialecka M, Novoa A, Carapuco M, van Nes J, de Graaff W, Duluc I, Freund JN, Beck F, Mallo M, Deschamps J. Cdx and Hox genes differentially regulate posterior axial growth in mammalian embryos. Dev Cell. 2009 Oct;17(4):516-526. doi: 10.1016/j.devcel.2009.08.010.

[47] Monie IW, Khemmani M. Absent and abnormal umbilical arteries. Teratology. 1973 Apr;7(2):135-141. doi: 10.1002/tera.1420070204.

[48] De Marco P, Merello E, Mascelli S, Raso A, Santamaria A, Ottaviano C, Calevo MG, Cama A, Capra V. Mutational screening of the CYP26A1 gene in patients with caudal regression syndrome. Birth Defects Res A Clin Mol Teratol. 2006 Feb;76(2):86-95. doi: 10.1002/bdra.20225.

[49] Nengyin Sheng, Zhihui Xie, Chen Wang, Ge Bai, Kejing Zhang, Qingqing Zhu, Jianguo Song, Francois Guillemot, Ye-Guang Chen, Anning Lin, and Naihe Jing. Retinoic acid regulates bone morphogenic protein signal duration by promoting the degradation of phosphorylated Smad1 PNAS November 2, 2010 107 (44) 18886-188891; https://doi.org/10.1073/pnas.1009244107

[50] Hoffman LM, Garcha K, Karamboulas K, et al. BMP action in skeletogenesis involves attenuation of retinoid signaling. J Cell Biol. 2006;174(1):101-113. doi:10.1083/jcb.200604150

Chapter 10

Wiskott-Aldrich Syndrome

Saeed Sepehrnia

Abstract

The Wiskott-Aldrich syndrome (WAS) could be a rare X-linked primary immunodeficiency disorder characterized by recurrent infections, eczema, and bleeding following thrombocytopenia. Despite the rarity of this syndrome, today our understanding of the cellular and molecular basis of the pathogenesis of this disease has increased and it's well established that this disorder encompasses a wide range of clinical disorders including immunodeficiency, atopy, autoimmunity, and cancer. Wiskott–Aldrich Syndrome protein (WASP) mutations are located throughout the gene and inhibit or regulate the conventional function of WASP. Thus classic WAS occurs when WASP is absent, X-linked thrombocytopenia when mutated WASP is expressed, and X-linked neutropenia when missense mutations occur within the Cdc42-binding site. Developments within the use of diagnostic tools, supportive care, and advances in allogeneic hematopoietic cell transplantation have remarkably reduced the mortality related to this disorder. Besides, gene therapy has provided optimistic perspectives on the treatment approaches of those patients.

Keywords: Primary immunodeficiency, Wiskott-Aldrich Syndrome, X-linked thrombocytopenia, X-linked neutropenia, Hematopoietic cell transplantation, Gene therapy

1. Introduction

Wiskott-Aldrich syndrome (WAS) is a rare X-linked recessive disorder that is defined in three categories consist of severe immunodeficiency, thrombocytopenia, and eczema. Moreover, high frequency of lymphomas (mainly B cell origin), susceptibility to autoimmune diseases, recurrent infections, variable defects in B and T-lymphocyte function ,and bloody diarrhea are classified as other symptoms [1]. A genetic mutation in the WAS gene encoding Wiskott-Aldrich syndrome protein (WASp) affecting the immune system and leads to immunodeficiency. There is broad-spectrum depending on gene mutations ranging from severe phenotype (Classic WAS) to mild ones which consist of X-linked thrombocytopenia (XLT) and X-linked neutropenia (XLN) [2, 3]. The purpose of this chapter is to discuss WASp function, clinical and pathological manifestations associated with mutations of the WAS gene as well as a comprehensive review of WAS clinical diagnosis and treatment methods and their complications. In addition, a range of therapeutic approaches will be discussed, including the use of allogeneic hematopoietic stem cell (HCT) transplantation and gene therapy.

 IntechOpen

2. Etiology

Wiskott-Aldrich syndrome is caused by mutations of the gene WASP (WAS protein) involved in actin polymerization, cellular signaling, cytoskeletal rearrangement, and immunological synapse formation. These mutations cause changes in protein structure by different mechanisms and lead to different phenotypes of this disease [1, 2].

3. Epidemiology

It has been estimated that one in every 100,000 male births leads to Wiskott-Aldrich Syndrome, and in families without a history of the disease, the average age of diagnosis is 24 months [4, 5]. In the United States, the estimated incidence of WAS is one in 250,000 male births. Whereas WAS still is a rare disease, it is more common than other immunodeficiency syndromes such as hyper-IgM syndrome or SCID which have an estimated incidence of about one in 1,000,000 live births. It is thought WAS accounts for 1.2% of all inherited immunodeficiencies in the United States [6]. Up to now, the impact of any ethnic or geographical factor on its incidence has not been reported. Because of misdiagnosing of mild cases, these conditions may be presumed as idiopathic thrombocytopenia purpura [4].

4. Clinical and pathological manifestations

4.1 Incidence and clinical phenotypes

The incidence of the classic WAS phenotype has been estimated at 1 in 100,000 individuals. Based on clinical manifestations, WAS-XLT commonly is presented at birth and consists of bruising, bloody diarrhea as well as petechiae. Excessive bleeding consequent circumcision can be considered as an early diagnostic sign. Furthermore, during infancy and childhood, eczema is a frequent manifestation of patients with classic WAS. Small platelet and thrombocytopenia are the most reliable finding in WAS and XLT phenotype. A variety of infections such as bacterial pneumonia, skin infections as well as otitis media with drainage of mucoid material are common complaints. XLT patients are less likely to have problems such as eczema and infection and oftentimes are misdiagnosed with Idiopathic thrombocytopenic purpura (ITP). Patients with X-linked neutropenia caused by missense mutations in the Cdc42-binding domain are affected at birth. However, their symptoms do not resemble those of classic WAS or XLT. Through a simple scoring system, we have delineated different clinical phenotypes (**Table 1**).

4.2 Thrombocytopenia

It is the foremost common finding present at the time of diagnosis and appears at birth. However, the severity is variable. Approximately 50% of patients with WAS will have severe Thrombocytopenia with a platelet count<20,000/μL. Furthermore, the severity of Thrombocytopenia has a close relationship with bleeding. Bleeding events occur in >80% of WAS patients. These bleeding complications consist of non-life-threatening (e.g, ecchymosis, hematemesis, epistaxis, petechiae, etc.) and or life-threatening (eg, gastrointestinal and intracranial hemorrhage). life-threatening bleeding happens in 30% of patients. However, intracranial hemorrhage occurs in exactly 2% of cases. Although Megakaryocyte numbers are typically

	WAS	XLT	IXLT	XLN
Phenotype				
Thrombocytopenia	+	+	(+)	-
Small platelets	+	+	+	-
Eczema	+/++/+++	-/+	-	-
Immune deficiency	+/++	-/(+)	-	-
Infections			-	+*
Autoimmunity and/or malignancies	Frequent	Possible	-	-
Congenital neutropenia	-	-	-	+
Disease scores	3,4, or 5	1,2, or (5)†	<1	0
WASP expression	Absent or truncated	Present, reduced quantity	Present, normal quantity	Present
Treatment				
IVIG	Yes	No (with exceptions)	No	No
HSCT	Yes at an early age	Might be considered if there is a sibling donor	No	?
Splenectomy	No	Might be considered‡	No	No

IXLT, Intermittent XLT; HSCT, hematopoietic stem cell transplantation; XLN, X-linked neutropenia.
Scoring system: -/ (+), absent or mild; -/+, intermittent thrombocytopenia; (+), mild, transient eczema or mild, infrequent infections not resulting in sequelae; +, thrombocytopenia, persistent but therapy-responsive eczema, and recurrent infections requiring antibiotics and often IVIG prophylaxis; ++, eczema that is difficult to control and severe, life-threatening infections.
**Infections typical for neutropenia.*
†Patients with XLT with a score of 1 or 2 might progress to a score of 5. Incidence of autoimmunity and malignancies are less in XLT than in WAS.
‡Splenectomy results in increased platelet numbers and reduced bleeding but causes a marked increase in sepsis, requiring continuous antibiotic prophylaxis.

Table 1.
Clinical phenotype associated with mutations of the WASP gene [acc.19].

normal in patients with WAS, platelet formation changes into abnormal. It should be noted that the role of anti-platelet antibodies in platelet destruction and maintaining thrombocytopenia isn't negligible. As an example, a group of infants less than or equal to 2 years of age may present with "severe refractory thrombocytopenia," probably due to antiplatelet autoantibody. Moreover, disruptions in platelet function as well as remarkable and may give rise to bleeding and diminished platelet survival. WASp plays a key role in the process of platelet formation, activation, and associated cytoskeletal remodeling so that failure in platelet production and function is attributed to it. Increased expression of phosphatidylserine on the surface of circulating platelets has been interpreted as a sign of increased phagocytosis and destruction of platelets within the spleen in WAS/XLT. Whereas, decreased platelet production resulting from ineffective thrombocytopoiesis [6–8].

4.3 Immunodeficiency

Abnormalities in immune system function (i.e., cell-mediated, humoral, and innate immunity) among patients with WAS give rise to vulnerability to a wide range of infections pathogens. Nevertheless, infectious complications as a primary manifestation are not frequent (<5% of cases). Opportunistic infections such as molluscum contagiosum infections, extensive candidiasis, aspergillosis, and

Pneumocystis jirovecii may also present in WAS patients [5, 9, 10]. Patients with WAS can suffer from severe and disseminated forms of herpes simplex virus I or II (6% of cases) and varicella (3% of cases) as the most widespread pathogens. In 10% of cases, Invasive yeast and fungal infections have been reported. Generally, Sinopulmonary infections are classified as the most common infectious manifestation before diagnosis, including otitis media (64% of cases) and pneumonia. Other severe infectious complications are less likely to occur, such as sepsis (7% of cases) and meningitis (4% of cases) [6]. However, Viral infections caused by pathogens such as EBV, CMV, and HPV can be extremely severe.

In WAS patients, both quantitative and qualitative defects in T cells are manifested T cell lymphopenia as a common disorder is seen more in naive T lymphocytes than in memory cells and CD8. This event may result from increased apoptosis and can appear from early life and subsequently affect thymic output [11, 12]. More often than not, WAS patients have got an absolute lymphocyte count >1,000/µL, demonstrating the lack of the profound lymphopenia seen in other primary immunodeficiency disorders. Of note, an absolute lymphocyte count of <1,000/µL was presented in only 22% of cases. Processes such as T-cell activation, chemotaxis, and cytokine secretion are disturbed in patients with WAS [13, 14].

Humoral immunodeficiency is another characteristic of WAS patients. High serum levels of IgE and Low levels of IgM, IgG, and IgA are observed [5]. Abnormal isohemagglutinin titers (84% of cases) and insufficient vaccine responses to protein (e.g., 62% with abnormal responses to tetanus vaccine), polysaccharide (e.g., 69% with abnormal responses to the pneumococcal vaccine), and conjugate vaccines (e.g., 66% with abnormal responses to Hib (*Haemophilus influenzae* type b) conjugate vaccines, implies that antibody responses may be abnormal [5]. Malfunctions in T-cell mentioned earlier may disturb the maturation and differentiation of B cells into antibody-producing cells and memory cells [15].

Perturbations in the components of the innate immune response may also be present. The number of Natural Killer (NK) cells can be in the normal range or increase. Nonetheless, most of the time, NK cell function is abnormal, including Immunological synapse formation, stimulation of secretory granules, and consequent cytolytic activity [16]. Despite the normal numbers of neutrophils, monocytes, and other phagocytes, functional abnormalities may be present in WAS patients. Chemotaxis, adhesion/arrest function, DC motility, the initiation of degranulation, the formation of a functional respiratory burst, and antibody-mediated phagocytosis are more likely to be impaired [17, 18].

4.4 Eczema

Eczema is one of the specific findings that essentially leads to the differentiation of WAS from ITP. Skin manifestations resemble acute or chronic eczema in appearance and distribution. Eczema of varying severity occurs in approximately 50% of WAS patients during the first year of life and resembles classic atopic dermatitis. [2]. According to a large cohort study, 81% of WAS patients are classified into mild or severe, transient, or consistent, depending on their eczema history. In severe form, eczema resists therapy and lasts into adulthood. Based on some statistical evidence, Patients with XLT either have mild transient eczema or do not have any of these symptoms. It has been hypothesized that defects in the chemotaxis of DC and Langerhans cells, which are responsible for presenting specific (probably bacterial) antigens to T lymphocytes, cause eczema. Eczema is more complex in families with a history of atopic disease because the findings suggest that genes involved in allergic processes may have a modulatory effect [19, 20].

4.5 Autoimmune manifestations

According to collected data, autoimmunity is a frequent occurrence in patients with WAS. Reports indicate that 40% of patients with WAS develop autoimmunity and that many patients show various forms of autoimmune disease. A study affirms that Two-thirds of WAS patients who show autoimmune manifestations develop multiple autoimmune disorders. Autoimmune hemolytic anemia (14%), vasculitis (13%), renal disease (12%), and chronic arthritis (10%) are the most common manifestations of autoimmunity in WAS [5]. Autoimmunity in WAS may be due to the formation of autoantibodies or the presence of autoreactive T cell clones. Moreover, disorders in the homeostasis of regulatory T cells and B lymphocytes can lead to autoimmune disorders in WAS patients. Generally, the incidence of autoimmune disease is lower XLT than in classic WAS. A broad spectrum of autoantibodies has been detected both in classic WAS and in XLT. For example, high serum IgM levels are a risk factor for the development of autoimmune disease and early death [21, 22].

4.6 Malignancies

Malignancies can occur during childhood, whereas it is most likely to appear in adolescents and young adults with the classic WAS phenotype. B cell lymphoma (often Epstein-Barr virus-positive), leukemia, myelodysplasia, and myeloproliferative disorders are among the most frequent malignancies [6, 23]. Based on a cohort study, 13% of patients with WAS developed malignancy at a mean age of 9.5 years, and only 1 of 21 patients who developed a malignancy was alive more than two years after diagnosis [5]. The incidence of malignancies in patients with XLT phenotype is ambiguous and less than in classic WAS. In WAS patients, the prevalence of non-Hodgkin lymphoma (NHL) is more common than Hodgkin lymphoma (HL) [23]. Several NHL, such as Burkitt lymphoma and lymphoblastic lymphoma, have been reported rarely among patients with WAS. The aggressive nature of malignancy in the WAS patients represents a poor prognosis, as data demonstrate 95% mortality among patients. It should be noted that genetic susceptibility due to malfunction of WASp, associated abnormalities in tumor-surveillance mechanisms (e.g., impairment in NK cell cytotoxicity), and environmental factors (e.g., Epstein–Barr virus [EBV]) are significant components that increase the risk of malignancy in patients with WAS [24].

4.7 X-linked thrombocytopenia and X-linked neutropenia

XLT is assumed as congenital thrombocytopenia that is sometimes intermittent (IXLT). In such cases, the eczema is usually mild. Generally, XLT patients have got a benign disease as well as excellent survival in contrast to patients with WAS. serious hemorrhage in 13.9%, life-threatening infections in 6.9%, autoimmunity in 12.1%, and malignancy in 5.2% of XLT patients at median ages of 4.9 years, 24.8 years, 12.2 years, and 34.0 years, respectively. Every male with thrombocytopenia and small platelets should be evaluated for WASp expression and WAS gene mutations [25].

XLN presents mainly as a congenital and severe form of neutropenia. Unlike WAS, infectious complications due to T-cell immunodeficiency are absent. Impairments in immune function are similar to those explained for WAS. Nevertheless, Decreased NK cell count is a valid finding in XLN patients. Also, a slight decrease in platelet count has been reported. The potential risk of myelodysplastic syndrome and chronic myelocytic leukemia exists, which needs regular surveillance [26].

5. Histopathology

Gradually, the cellular elements in the thymus and lymphatic organs begin to disappear [27]. Depletion of small lymphocytes from T cell areas, reticular cell stroma protrusions, and the presence of abnormal plasma cells, often associated with extracellular plasmacytosis and hematopoiesis, are consistently seen in the lymph nodes and spleen of patients with WAS [28]. In a study of spleens obtained from WAS patients undergoing splenectomy, gradual depletion of the marginal zone involving B cells was also observed [29]. It may be possible to justify the disruption of the antibody response to selected polysaccharides and protein antigens by examining these histological abnormalities.

6. Diagnosis

WAS is an X-linked disease presented in males, with a lack of clinical symptoms in female carriers. Also, a deleterious mutation of the paternally derived X chromosome and inactivation of the maternally derived X chromosome lead to WAS in females, which is rare [30]. The diagnosis of WAS should be conducted in any male appearing with thrombocytopenia, eczema, recurrent respiratory infections, autoimmunity, etc (**Table 2**). Clinical findings may or may not be present during the course of the disease. Therefore, Because of evolution in clinical, physical, and laboratory findings, there is a dire need for Reassessment over time [5].

Physical exams	
Rash	Eczema
Bleeding	Petechiae, ecchymoses
Past medical history	
Rash	Eczema
Bleeding	Mucosal bleeding (easy bruising, epistaxis, hematochezia, hematuria) or intracranial hemorrhage
Infections	Recurrent or severe sinopulmonary infections, viral infections, fungal infections, or opportunistic infections
Autoimmunity	Cytopenias, vasculitis, inflammatory bowel disease, arthritis, renal disease
Malignancy	Lymphoma
Family history	
X-linked disorder	Every generation affected; predominant male susceptibility
Laboratory exam	
Complete blood cell count	Anemia, microcytosis, thrombocytopenia, low mean platelet volume
Peripheral blood smear	Microthrombocytes
Serum IgG, IgA, IgM, and IgE	Low lgG, lgA, lgM; high lgE
Isohemagglutinin and vaccine titers	Abnormal isohemagglutinin titers and diminished vaccine responses to protein, polysaccharide, and conjugate vaccines
Lymphocyte subsets and mitogen responses	T-cell lymphopenia and abnormal proliferative responses to mitogens

Abbreviation: Ig, immunoglobulin.

Table 2.
Clues to diagnosis of Wiskott–Aldrich syndrome [acc.6].

Despite the broad spectrum of clinical features, there is a considerable association between genotype and phenotype [9]. Mutations give rise to absent WASp expression and residual WASp expression are related to classic syndrome and XLT phenotype respectively. Also, gain-of-function mutations in the WAS gene result in XLN [31].

Interestingly, the patient phenotype may undergo genetic reversion, a selective advantage that can produce hematopoietic cells with normal WASP expression [32]. Screening for WASP mutations can be accomplished by flow cytometry using a suitable anti-WASP antibody. With this method, patients with expression of mutated WASP are likely to be miss. However, scrutinizing Sequence analysis of the WASP gene is essential to establishing a final diagnosis. It is speculated that the combination of two methods may help to estimate the severity of the disease [7].

7. Prognosis

The prognosis of X-linked thrombocytopenia has got a favorable position with a life expectancy close to the normal population [25]. Whereas, Classic Wiskott Aldrich syndrome has a poor prognosis with decreased life expectancy, which can be attributed to recurrent infections, autoimmune disease, and malignancy. Hemorrhage is the main cause of death in these patients [5].

As soon as a diagnosis is confirmed, the patient should be monitored and evaluated in a center with expertise in the management of WAS. It is clear that Without appropriate care and intervention, morbidity and mortality will not be unexpected. In a retrospective research, it has been demonstrated 36% of patients with WAS experienced non-HSCT-associated deaths at a mean age of 8 years. Mainly, these deaths are caused by infection (44%), bleeding (23%), and malignancy (26%) [5].

Many centers that provide look after WAS patients have multifaceted approaches to the care of patients and affected members of family, including genetic counseling, psychosocial support services, and subspecialist support, like transplant immunology, hematology/oncology, communicable disease, and important care. Of note, with appropriate care and timely intervention, WAS patients have an interesting prognosis. As an example, long-term survival following the utilization of allogeneic HSCT is >80% [31].

8. Treatment

Basically, the treatment of Wiskott-Aldrich syndrome depends on supportive care which includes Broad-spectrum antibiotics for bacterial, fungal, or viral infections. Furthermore, platelet supplement, Topical steroids and prevent bleeding, are other treatments. However, a series of controversial treatment are as follows (**Table 3**).

8.1 Intravenous Immunoglobulin therapy

Intravenous immunoglobulin (IVIG) therapy in WAS and XLT patients with significant antibody deficiency has led to efficient results. IVIG should be administered at physiological doses to patients with recurrent infectious complications and low levels of immunoglobulin or abnormal antibody responses [6]. Because of the increasing catabolic rate observed in WAS patients, the dose is higher than other immunodeficiency diseases. Since these people are more likely to bleed, a Subcutaneous injection of the IVIG is recommended [33]. In the presence of autoimmune manifestations, at least intermittently, immunosuppressive therapy may be required.

Conventional treatments
Broad-spectrum antibiotics
platelet supplement
Topical steroids
Potential treatments
Intravenous immunoglobulin (IVIG)
Splenectomy
Eltrombopag
Immunosuppressive treatment
Hematopoietic cell transplantation (HCT)
Gene therapy

Table 3.
Therapeutic approaches in Wiskott-Aldrich syndrome.

8.2 Splenectomy

Splenectomy is used to slow down the process of thrombocytopenia and stop bleeding by increasing the number of circulating platelets. Patients undergoing splenectomy consume antibiotics for life also they are highly vulnerable to septicemia [2]. Splenectomy is not recommended for people who are going to have Hematopoietic cell transplantation (HCT) in the future because it increases the risk of significant infectious complications [6, 7].

8.3 Eltrombopag

It is an oral thrombopoietin receptor agonist approved for the treatment of ITP which is claimed to probably effective in preventing bleeding in patients with WAS waiting for HCT [34].

8.4 Immunosuppressive treatment

More often than not, prescribing immunosuppressive drugs (e.g., cyclophosphamide, azathioprine) is necessary for the autoimmune phenomenon [35]. Most of the time, Monoclonal antibody rituximab may be effective in cytopenias due to Autoimmune disorders and it should be noted that the aforementioned antibody partly is harmless for patients already receiving therapy with IVIG [2].

8.5 Hematopoietic cell transplantation (HCT)

In a way, it can be said HCT is a unique treatment for patients with human leukocyte antigen (HLA)-matched family or unrelated donors (URDs) or relatively matched cord-blood donors [2].

Newly, according to haploidentical donors, novel graft manipulation approaches that alleviate the risk of graft versus host disease (GVHD) and elevate the possibility of successful engraftment, as well as immune system reconstruction, have yielded promising results in patients with WAS [36]. Reports indicate that The older you are at the time of transplantation, the lower your chances of survival, with different studies indicating different thresholds (e.g., age < 2 or < 5 years) [37, 38].

Despite the high chance of survival after transplantation for WASP patients, these patients are still at risk for post-HCT complications, among which autoimmunity is a prominent problem. It has been reported that up to 55% of transplanted WAS subjects develop autoimmune manifestations, mostly involving antibody-mediated cytopenias [39–42]. Although the risk of death in HCT is inevitable, according to a pervasive theory, patients with the severe phenotype (clinical score 3 and above) should be transplanted. For WAS patients with milder clinical manifestations, there is no unanimous opinion, and the decision to continue HCT is made on a case-by-case basis. The results of the decision-making process were the result of a retrospective study of the outcome of HCT in a group of 24 milder patients between 1990 and 2011 who were transplanted at various centers around the world. This study indicated a survival rate of 83% in the absence of long-term complications [43].

8.6 Gene therapy

Approximately, HCT from HLA-identical family donors is available to less than 20% of transplanted WAS patients [40, 43]. Although this approach is evolving [36], the increasing complications and mortality of HCT from mismatched donors [40] provide the idea for research into alternative kinds of gene therapy. Gene therapy prepares an enormous of potential advantages over allogeneic HCT, including availability for all patients, reduce transplant rejection risks and prevent GVHD risks, which successively eliminates the necessity to follow special diets and take immunosuppressive drugs. For the primary time in Germany, gene therapy for WAS was accomplished in a clinical trial, during which a gamma-retroviral vector was used to correct CD34+ cells from ten WAS patients. As a result of this investigation, nine of those individuals showed a significant increase in platelet count and rehabilitation of immune responses. But unfortunately, seven patients developed acute leukemia related to vector integration-mediated activation of the LMO2, MDS1, or MN1 genes [44, 45].

A High and unacceptable rate of cancer incidence in gamma-retroviral gene therapy paves the way for the implementation of clinical gene transfer protocols using HIV 1-based constructs [46]. Currently, a series of clinical trials in Europe and the U.S. using these lentiviral vectors have yielded encouraging preliminary results In 2010, when the primary trial was launched, Italian investigators have treated a minimum of 10 patients (F. Ferrua, personal communication, Barcelona, September 2016) with infusions of gene-corrected bone marrow and/or mobilized peripheral blood CD34+ hematopoietic cells. Then, they controlled the possible side effects by prescribing busulfan, fludarabine, and rituximab. The results of the first three patients ≥ 1 year after gene therapy indicated an improvement in platelet count and immune cell function additionally a reduction in severe infections and an improvement in eczema [47, 48]. The results of studies on the effect of gene therapy on defective B cells in patients with WAS indicate that standard distribution of bone marrow and peripheral blood cell subsets, in treated patients, is achievable. most significantly, a serious decrease within the abundance of naive B cells producing reactive antibodies, which are involved in improving the quantity of circulating antibodies altogether treated patients, was observed [49, 50].

The second investigation, conducted in London and Paris, using the identical lentiviral vector and a uniform reparative chemotherapy regimen, 6 out of seven patients treated also demonstrated improvement of immune function and clinical manifestations during 6–42 months of follow-up. Furthermore, during this study, no vector-mediated clonal expansions have occurred [51]. Of note, although in both trials the duration of bleeding was significantly reduced in number and severity, and also the treated patients not needed blood transfusions and thrombopoietin receptor agonists, platelet counts failed to normal in either trial that it isn't clear. The third trial of WAS gene therapy supported lentiviral vector has recently begun in Boston, USA [52]. At the identical time, other US researchers have developed an alternate lentiviral vector with a stronger WASp expression that's being developed for future clinical applications [53, 54].

In line with current studies, it will be acknowledged that gene modification by lentivirus from autologous hematopoietic ancestors can have significant benefits for patients undergoing treatment and be considered as treatment options for WAS. However, more comprehensive studies are needed to ascertain whether this type of gene therapy could also be a definitive treatment for patients.

9. Conclusions

Despite the rarity of WAS, extensive progress has been made in understanding its pathophysiological foundations, but it is still necessary to establish multifaceted management to assess various aspects of the disease. It is worth noting that health care centers have been pioneers in the diagnosis and management of WAS. Significant advances in allogeneic HCT and its valuable long-term results have made it a viable treatment option for most patients with WAS. For severe manifestations, for example, definitive treatment with HCT is recommended. However, even in a clear clinical situation, HCT may not be available due to the patient's geographical location or socioeconomic status, so supportive care measures should be taken promptly. The same is true of WAS cases with milder clinical manifestations; In fact, more emphasis is placed on accepting the potential risks of treatment in proportion to the severity of the disease manifestations.

Gene therapy is a potential treatment solution for WAS patients with severe and even mild phenotypes. In this regard, the emergence of advances in the use of gene editing technology creates a cautious optimism. However, the financial and geographical problems for patients with limited access to gene therapy options need to be addressed.

Acknowledgements

This work has been financed by Intechopen publications and I would like to thank Dr. Hridayesh Prakash as well as dear Ms. Marijana Francetic for closed cooperation and sincere support. I thank Dr. David Buchbinder from Children's Hospital of Orange Country and Dr. Hans D. Ochs from the University of Washington for their sincere cooperation and support.

Conflict of interest

The authors declare no conflict of interest.

Thanks

I am grateful to my family, Dr. Tooba Ghazanfari from Shahed University, Dr. Farideh Talebi from Shahed University, Dr. Mohammad Mahmoudi from Iran

University, Dr. Hamed Afkhami from Shahed University, Dr. Roghaye Afifirad from Tehran University as well as MSc. Sayed Esmaeil Ahmadi from Iran University and also Mr. Mohammad Hosein Siahkary from Qom Islamic Azad University.

Abbreviations used

HCT	Hematopoietic stem Cell
ITP	Idiopathic thrombocytopenic purpura
IVIG	Intravenous immunoglobulin
WAS	Wiskott-Aldrich syndrome
WASp	Wiskott-Aldrich syndrome protein
XLT	X-linked thrombocytopenia
XLN	X-linked neutropenia

Author details

Saeed Sepehrnia
Department of Immunology, School of Medicine, Shahed University, Tehran, Iran

*Address all correspondence to: sepehrniasaeed@gmail.com

References

[1] Pituch-Noworolska, A., *Autoimmunity in Children with Primary Immunodeficiency – Diagnosis, Management and Therapy*, in *Autoimmunity - Pathogenesis, Clinical Aspects and Therapy of Specific Autoimmune Diseases*. 2015.

[2] Malik, M.A. and M. Masab, *Wiskott-Aldrich Syndrome*, in *StatPearls*. 2020, StatPearls Publishing Copyright © 2020, StatPearls Publishing LLC.: Treasure Island (FL).

[3] Bosticardo, M., et al., *Recent advances in understanding the pathophysiology of Wiskott-Aldrich syndrome.* Blood, 2009. **113**(25): p. 6288-95.

[4] Stray-Pedersen, A., T.G. Abrahamsen, and S.S. Frøland, *Primary immunodeficiency diseases in Norway.* J Clin Immunol, 2000. **20**(6): p. 477-85.

[5] Sullivan, K.E., et al., *A multiinstitutional survey of the Wiskott-Aldrich syndrome.* J Pediatr, 1994. **125** (6 Pt 1): p. 876-85.

[6] Buchbinder, D., D.J. Nugent, and A.H. Fillipovich, *Wiskott-Aldrich syndrome: diagnosis, current management, and emerging treatments.* Appl Clin Genet, 2014. **7**: p. 55-66.

[7] Ochs, H.D., et al., *Wiskott-Aldrich syndrome: diagnosis, clinical and laboratory manifestations, and treatment.* Biol Blood Marrow Transplant, 2009. **15**(1 Suppl): p. 84-90.

[8] Mahlaoui, N., et al., *Characteristics and outcome of early-onset, severe forms of Wiskott-Aldrich syndrome.* Blood, 2013. **121**(9): p. 1510-6.

[9] Imai, K., et al., *Clinical course of patients with WASP gene mutations.* Blood, 2004. **103**(2): p. 456-64.

[10] Zhang, Z.Y., et al., *Analysis of clinical and molecular characteristics of Wiskott-Aldrich syndrome in 24 patients from 23 unrelated Chinese families.* Pediatr Allergy Immunol, 2010. **21**(3): p. 522-32.

[11] Park, J.Y., et al., *Early deficit of lymphocytes in Wiskott-Aldrich syndrome: possible role of WASP in human lymphocyte maturation.* Clin Exp Immunol, 2004. **136**(1): p. 104-10.

[12] Rawlings, S.L., et al., *Spontaneous apoptosis in lymphocytes from patients with Wiskott-Aldrich syndrome: correlation of accelerated cell death and attenuated bcl-2 expression.* Blood, 1999. **94**(11): p. 3872-82.

[13] Zhang, J., et al., *Antigen receptor-induced activation and cytoskeletal rearrangement are impaired in Wiskott-Aldrich syndrome protein-deficient lymphocytes.* J Exp Med, 1999. **190**(9): p. 1329-42.

[14] Calvez, R., et al., *The Wiskott-Aldrich syndrome protein permits assembly of a focused immunological synapse enabling sustained T-cell receptor signaling.* Haematologica, 2011. **96**(10): p. 1415-23.

[15] Recher, M., et al., *B cell-intrinsic deficiency of the Wiskott-Aldrich syndrome protein (WASp) causes severe abnormalities of the peripheral B-cell compartment in mice.* Blood, 2012. **119**(12): p. 2819-28.

[16] Orange, J.S., et al., *Wiskott-Aldrich syndrome protein is required for NK cell cytotoxicity and colocalizes with actin to NK cell-activating immunologic synapses.* Proc Natl Acad Sci U S A, 2002. **99**(17): p. 11351-6.

[17] Kumar, S., et al., *Cdc42 regulates neutrophil migration via crosstalk between*

WASp, CD11b, and microtubules. Blood, 2012. **120**(17): p. 3563-74.

[18] Dart, A.E., et al., *Nck and Cdc42 co-operate to recruit N-WASP to promote FcγR-mediated phagocytosis.* Journal of cell science, 2012. **125**(Pt 12): p. 2825-2830.

[19] Ochs, H.D. and A.J. Thrasher, *The Wiskott-Aldrich syndrome.* J Allergy Clin Immunol, 2006. **117**(4): p. 725-38; quiz 739.

[20] Thrasher, A.J., et al., *Is Wiskott--Aldrich syndrome a cell trafficking disorder?* Immunol Today, 1998. **19**(12): p. 537-9.

[21] Shcherbina, A., F.S. Rosen, and E. Remold-O'Donnell, *Pathological events in platelets of Wiskott-Aldrich syndrome patients.* Br J Haematol, 1999. **106**(4): p. 875-83.

[22] Dupuis-Girod, S., et al., *Autoimmunity in Wiskott-Aldrich syndrome: risk factors, clinical features, and outcome in a single-center cohort of 55 patients.* Pediatrics, 2003. **111**(5 Pt 1): p. e622-7.

[23] Cotelingam, J.D., et al., *Malignant lymphoma in patients with the Wiskott-Aldrich syndrome.* Cancer Invest, 1985. **3**(6): p. 515-22.

[24] Kersey, J.H., R.S. Shapiro, and A.H. Filipovich, *Relationship of immunodeficiency to lymphoid malignancy.* Pediatr Infect Dis J, 1988. 7(5 Suppl): p. S10-2.

[25] Albert, M.H., et al., *X-linked thrombocytopenia (XLT) due to WAS mutations: clinical characteristics, long-term outcome, and treatment options.* Blood, 2010. **115**(16): p. 3231-8.

[26] Devriendt, K., et al., *Constitutively activating mutation in WASP causes X-linked severe congenital neutropenia.* Nat Genet, 2001. **27**(3): p. 313-7.

[27] Cooper, M.D., et al., *Wiskott-Aldrich syndrome. An immunologic deficiency disease involving the afferent limb of immunity.* Am J Med, 1968. **44**(4): p. 499-513.

[28] Snover, D.C., et al., *Wiskott-Aldrich syndrome: histopathologic findings in the lymph nodes and spleens of 15 patients.* Hum Pathol, 1981. **12**(9): p. 821-31.

[29] Vermi, W., et al., *The spleen in the Wiskott-Aldrich syndrome: histopathologic abnormalities of the white pulp correlate with the clinical phenotype of the disease.* Am J Surg Pathol, 1999. **23**(2): p. 182-91.

[30] Boonyawat, B., et al., *Combined de-novo mutation and non-random X-chromosome inactivation causing Wiskott-Aldrich syndrome in a female with thrombocytopenia.* J Clin Immunol, 2013. **33**(7): p. 1150-5.

[31] Massaad, M.J., N. Ramesh, and R.S. Geha, *Wiskott-Aldrich syndrome: a comprehensive review.* Ann N Y Acad Sci, 2013. **1285**: p. 26-43.

[32] Davis, B.R. and F. Candotti, *Revertant somatic mosaicism in the Wiskott-Aldrich syndrome.* Immunol Res, 2009. **44**(1-3): p. 127-31.

[33] Blaese, R.M., et al., *Hypercatabolism of IgG, IgA, IgM, and albumin in the Wiskott-Aldrich syndrome. A unique disorder of serum protein metabolism.* J Clin Invest, 1971. **50**(11): p. 2331-8.

[34] Gerrits, A.J., et al., *Effects of eltrombopag on platelet count and platelet activation in Wiskott-Aldrich syndrome/X-linked thrombocytopenia.* Blood, 2015. **126**(11): p. 1367-78.

[35] Candotti, F., *Clinical Manifestations and Pathophysiological Mechanisms of the Wiskott-Aldrich Syndrome.* J Clin Immunol, 2018. **38**(1): p. 13-27.

[36] Kharya, G., et al., *Haploidentical T-cell alpha beta receptor and CD19-depleted stem cell transplant for Wiskott-Aldrich syndrome.* J Allergy Clin Immunol, 2014. **134**(5): p. 1199-201.

[37] Kobayashi, R., et al., *Outcome in patients with Wiskott-Aldrich syndrome following stem cell transplantation: an analysis of 57 patients in Japan.* Br J Haematol, 2006. **135**(3): p. 362-6.

[38] Shekhovtsova, Z., et al., *A risk factor analysis of outcomes after unrelated cord blood transplantation for children with Wiskott-Aldrich syndrome.* Haematologica, 2017. **102**(6): p. 1112-1119.

[39] Ozsahin, H., et al., *Long-term outcome following hematopoietic stem-cell transplantation in Wiskott-Aldrich syndrome: collaborative study of the European Society for Immunodeficiencies and European Group for Blood and Marrow Transplantation.* Blood, 2008. **111**(1): p. 439-45.

[40] Moratto, D., et al., *Long-term outcome and lineage-specific chimerism in 194 patients with Wiskott-Aldrich syndrome treated by hematopoietic cell transplantation in the period 1980-2009: an international collaborative study.* Blood, 2011. **118**(6): p. 1675-84.

[41] Friedrich, W., et al., *Results and long-term outcome in 39 patients with Wiskott-Aldrich syndrome transplanted from HLA-matched and -mismatched donors.* Immunol Res, 2009. **44**(1-3): p. 18-24.

[42] Shin, C.R., et al., *Outcomes following hematopoietic cell transplantation for Wiskott-Aldrich syndrome.* Bone Marrow Transplant, 2012. **47**(11): p. 1428-35.

[43] Oshima, K., et al., *Hematopoietic Stem Cell Transplantation for X-Linked Thrombocytopenia With Mutations in the WAS gene.* J Clin Immunol, 2015. **35**(1): p. 15-21.

[44] Boztug, K., et al., *Stem-cell gene therapy for the Wiskott-Aldrich syndrome.* N Engl J Med, 2010. **363**(20): p. 1918-27.

[45] Braun, C.J., et al., *Gene therapy for Wiskott-Aldrich syndrome--long-term efficacy and genotoxicity.* Sci Transl Med, 2014. **6**(227): p. 227ra33.

[46] Dupré, L., et al., *Lentiviral vector-mediated gene transfer in T cells from Wiskott-Aldrich syndrome patients leads to functional correction.* Mol Ther, 2004. **10**(5): p. 903-15.

[47] Aiuti, A., et al., *Lentiviral hematopoietic stem cell gene therapy in patients with Wiskott-Aldrich syndrome.* Science, 2013. **341**(6148): p. 1233151.

[48] Ferrua, F., et al., *Lentiviral haemopoietic stem/progenitor cell gene therapy for treatment of Wiskott-Aldrich syndrome: interim results of a non-randomised, open-label, phase 1/2 clinical study.* Lancet Haematol, 2019. **6**(5): p. e239-e253.

[49] Pala, F., et al., *Lentiviral-mediated gene therapy restores B cell tolerance in Wiskott-Aldrich syndrome patients.* J Clin Invest, 2015. **125**(10): p. 3941-51.

[50] Castiello, M.C., et al., *B-cell reconstitution after lentiviral vector-mediated gene therapy in patients with Wiskott-Aldrich syndrome.* J Allergy Clin Immunol, 2015. **136**(3): p. 692-702.e2.

[51] Hacein-Bey Abina, S., et al., *Outcomes following gene therapy in patients with severe Wiskott-Aldrich syndrome.* Jama, 2015. **313**(15): p. 1550-63.

[52] Kumar, S.R., et al., *Clinical development of gene therapy: results and lessons from recent successes.* Mol Ther Methods Clin Dev, 2016. **3**: p. 16034.

[53] Wielgosz, M.M., et al., *Generation of a lentiviral vector producer cell clone for*

human Wiskott-Aldrich syndrome gene therapy. Mol Ther Methods Clin Dev, 2015. **2**: p. 14063.

[54] Koldej, R.M., et al., *Comparison of insulators and promoters for expression of the Wiskott-Aldrich syndrome protein using lentiviral vectors.* Hum Gene Ther Clin Dev, 2013. **24**(2): p. 77-85.

Chapter 11

Fetal Congenital Anomalies in Africa: Diagnostic and Management Challenges

Labaran Dayyabu Aliyu

Abstract

There is paucity of knowledge on the causes, diagnosis, management and prevention of fetal congenital anomalies in Africa. The chapter will highlight on the general causes and specific factors concerning congenital anomalies in Africa. The problems of diagnosis and management of congenital anomalies will be extensively discussed. There is also going to be a discussion on how fetal anomalies contribute to maternal and perinatal mortality and morbidity. Screening of congenital anomalies is another black point and will be discussed emphasizing on simple strategies applicable in resource constrained environment. A section will be dedicated on prevention of fetal congenital anomalies, particularly prevention of specific factors that increase the risk of fetal anomalies in Africa. Finally, there will also be discussion on collaborative care as a panacea in the management of fetal congenital fetal anomalies, including my experience in this area. Specific examples will be given to illustrate the utility of collaborative in resource limited countries.

Keywords: congenital, anomalies, diagnosis, management, Africa

1. Introduction

Congenital anomalies can be defined as structural or functional anomalies [e.g. metabolic disorders] that occur during intrauterine life and can be identified prenatally, at birth or sometimes may only be detected later in infancy, such as hearing defect [1]. Every year, an estimated 7.9 million children are born with serious birth defects of genetic or partly genetic origin [2]. Over 1 million more infants are born with serious birth defects of post-conception origin including those that result from maternal exposure to environmental agents [teratogens] such as alcohol, rubella, syphilis, and iodine deficiency that can harm the developing fetus [3]. Thus, an estimated 9 million infants–representing approximately 7% of births-are born annually with defect that may kill them or results in lifelong disability [4]. An estimated 270,000 newborns die during the first 28 days of life every year from congenital anomalies [5]. Ninety-four percent of children with birth defects are born in low-income countries and 95% of those who die as a result of birth defect also born there [6]. Accurate data of prevalence of congenital fetal anomalies rare. Data is usually obtained from registries of congenital anomalies and this indicates that congenital anomalies are seen in 2–3% of newborns which is similar to what is seen in the industrialized world [7]. Congenital anomalies account for 8–15% of perinatal deaths and 13–16% of neonatal deaths in India [8]. Reliable data from low-income countries on fetal anomalies like data from other health indices is difficult to come by and even where available it is generated from institutional studies rather than from population based studies. Yet most congenital anomalies and their severe consequences are seen in developing low-income. In developed countries common causes of perinatal and neonatal have been dealt with and congenital anomalies are now seen as causes of perinatal and neonatal death. In developing countries, the reverse is the case and this may be the reason why the contributions of congenital anomalies in perinatal and neonatal mortality is well appreciated. So the augment goes in view of the presence of commoner causes of perinatal and neonatal mortality it will not be appropriate to allocate resources trying to reduce mortality from congenital anomalies. The is question now is, can we wait to achieve health transition in which common causes of perinatal and neonatal mortality are eliminated before addressing the issue of congenital anomalies and their contribution to perinatal and neonatal mortality? The answer is no. What we fail to realize is in developing countries congenital anomalies indirectly contribute to maternal mortality. Imagine a situation where a pregnant woman goes in to labor with a fetus with an undiagnosed congenital anomaly that preclude vaginal delivery and as we know more than half of pregnant women in developing countries labor and deliver at home. In this scenario the labor be prolonged and with time obstructed, membranes would have ruptured, chorio-amnionitis would have set in and as consequence develop postpartum hemorrhage or puerperal sepsis and die or she develop ruptured uterus and die. Congenital fetal anomalies can lead to both perinatal and maternal mortality and morbidity.

In this chapter, I will discuss congenital anomalies, their causes, prenatal diagnosis, treatment and prevention with the peculiarities of the African environment in view.

2. Background

Multitudes of factors determine the overall quality of health and pregnancy safety and outcome in Africa.

Illiteracy and poverty are factors that directly or indirectly influence health and pregnancy outcome in Africa. In Sub-Saharan Africa proportion of workers living in extreme poverty is 57% [9]. Other factors are ignorance, superstitious beliefs, bad cultural practices and poorly developed health infrastructure. The rate malnutrition in Sub-Saharan Africa is 23% [10]. In many countries in Africa utilization of antenatal care services is low and it is at this critical time pregnant women are screened for various diseases and ultrasound screening is done for fetal anomalies. In Nigeria only 2/3 of pregnant women attend antenatal care and only 40% deliver under the care of skilled birth attendants [10]. Utilization of the antenatal and delivery services in other African countries is much lower. Sub-Saharan Africa has the lowest contraceptive prevalence 13% and highest unmet need for family planning 28% [10]. Infections and infestations, lack of immunization against diseases that may be harmful to pregnancy and exposure to various potentially harmful substances increase the risk for fetal anomalies in Sub-Saharan African women. Lack of reliable data collection and recording means that the prevalence of congenital anomalies in the region is based on institutional estimates not the actual numbers. This is the background to understanding the causes, management and prevention of congenital anomalies in Africa. It is with this background in mind that this chapter will discuss the causes, management and prevention of congenital anomalies in Africa.

3. Causes of congenital anomalies

Fifty percent of birth defects have no clear identifiable cause and in the other 50% there are factors that considered as the cause. The causes can be broadly classified in to 3:

- Pre-conception
- Post-conception
- Birth defects of unknown cause (**Table 1**)

The pre-conception causes of birth defects are those causes that have their origin before conception and are genetic or partly genetic in origin [Genes and Chromosomes]. They found in families and can be inherited e.g. Sickle cell disease. They can also be seen as isolated incident in a particular pregnancy. The post-conception causes of birth defects are those anomalies that arise after conception or but before parturition. The last category of birth defects are those whose cause is unknown. The prevalence of birth defects based on the cause as shown in the above table is a broad division based on what is found in developed countries. If population based studies are conducted the findings differ from those above. This is because in developed countries concerted efforts were made to reduce environmental exposure to teratogens in pregnancy, institute preconception care to optimize medical conditions before pregnancy, immunize against infections that may affect the fetus in utero, offer pregnancy termination to where anomalies are identified etc. These measures reduce the prevalence of congenital anomalies of genetic origin or environmentally induced. In Africa environmental factors may play a role in causing birth defects. These may be from diseases [Viral, Bacterial and Protozoan/parasitic] or from exposure to teratogens [Alcohol, Cigarette, Pesticides and traditional medications who chemical constituents are unknown]. There is a difference in the annual numbers of birth defects, annual deaths from birth defects and annual under-5 deaths between Low-income, Middle-income and High-income countries.

Cause	%
Pre-conception	
Chromosome disorders	6
Single gene disorders	7.5
Multifactorial	20–30
Subtotal	40
Post-conception	
Teratogens	7–8%
Intrauterine abnormalities	2
Subtotal	10
Unknown cause	50
Total	100

Ternpenny and Ellard [11].

Table 1.
Percentage of birth defects by cause in high income countries.

	Low-income countries	Middle-income countries	High-income countries	Total
Annual total birth defects (millions)	4.75 60%	2.64 34%	0.49 6%	7.9
Annual early deaths of birth defects (millions)	2.38 72%	0.79 24%	0.14 4%	3.3
Annual under-5 deaths (millions)	8.8 80%	1.8 16%	0.6 4%	11.2
MOD [3]; UNICEF 2001.				

Table 2.
Estimated numbers and percentage of annual total birth defects, early deaths due to birth defects, and under-5 deaths for low-, middle-, and high- income countries.

This is a demonstration of the impact of strategies put in place to control birth defects based on the level of economic development and investments made in health care infrastructure and health care provision and prevalence of modifiable risk factors for birth defects (**Table 2**).

4. Chromosomal abnormalities as cause of birth defects

These account for 6% of birth defects in developed countries in industrialized countries [11]. The most common example of is Down syndrome which is characterized by an extra chromosome and is also called trisomy 21. This is condition is now diagnosed early [Thickened Nuchal translucency, absent or hypoplastic nasal born etc.] and pregnancy can be terminated. Other defects in this category include Edward's syndrome and patau syndrome. Many infants are born with Down syndrome in Africa because early diagnosis and termination is not possible. Lots of resources are expended by families and communities on caring for the affected infants which constitute a burden for the family.

Single gene defects an estimated 7.5% of birth defects [11]. They are caused by alteration gene structure and more than 6000 single gene defects were described.

Environmental factors as cause of birth defects.

Intrauterine infections;

Bacterial infections: Example, Syphilis which in Sub-Saharan Africa is seen 6–16% of pregnant women [12]. Currently most countries in Africa screen for Syphilis during antenatal care and provide treatment for those affected. Those affected present with features of the disease in the first 3 months of life which include; Vesiculobullous eruptions or macular copper-colored rash on the palms and soles and popular lesions around the nose and mouth as well as petechial lesions.

Protozoan infection: Congenital toxoplasmosis occurs as a result of maternal vertical transmission to the fetus. It is a cause of severe fetal complications that may manifest in the early neonatal period but may manifest later and lead to life-long complications. Diagnosis is through laboratory tests, however ultrasound is helpful and can be used to assess prognosis. The classical triad of congenital Toxoplasmosis are; chorioretinitis, hydrocephalus and intracranial calcifications.

Viral infections: Many viral infections are implicated as cause of several birth defects. Some viral diseases present with non-specific clinical features and many other infections have similar features. In Africa with poor health infrastructure screening and diagnosis for most diseases is challenging. Thus viral diseases may affect pregnant women and cause fetal congenital anomalies which may not in the long run be linked to the actual cause. Some common viral infections associated with congenital birth defects include; **Zika** virus, **Cytomegalovirus**, **Rubella** virus.

Zika virus: Infection Zika virus was first recorded in East Africa in 50s [Uganda and Tanzania]. It has recently caused epidemic in the Americas and travel advisory was issued with regard to this infection. The virus causes microcephaly and other congenital abnormalities known as Zika syndrome. It is also associated with other pregnancy complications including preterm birth and miscarriages.

Cytomegalovirus [CMV]: Women with cytomegalovirus infection have 1 in 3 chance of transmitting the infection to their fetus through the placenta. Not all exposed babies present with disease or its complications. There some ultrasound features detectable during prenatal screening which include; microcephaly, intracranial calcifications, ventriculomegally, ascites, hepato-splenomegally, intestinal, periventricular or hepatic echo densities and fetal hydrops. One or more of these may indicate congenital CMV.

Rubella virus: Rubella is one the most dangerous viral infection that lead to serious complications in the fetus. Approximately 25% of infant born to mothers who contract rubella in the first trimester of pregnancy have congenital rubella syndrome [CRS] [4]. In countries with successful rubella immunization programs, CRS has been eliminated. In the remaining 50% of countries, more than 100,000 infants are born with CRS annually [13]. The common birth defects from CRS include; cataract, heart defects, low birth weight, skin rash at birth, deafness, and intellectual disabilities. Others include glaucoma, brain damage, thyroid and other hormone problems.

5. Drugs and birth defects

Drugs of various types are known to cause congenital birth defects. In Africa most drugs are bought over the counter as there are strict regulations. Because of poverty low medications are preferred by patients and such drugs may have potential side effects including embryo toxicity. As most pregnancies are unplanned women on treatment for some medical conditions may become pregnant and continue taking treatment with drugs that are teratogenic to the fetus. As antenatal care patronage is low women on such drugs may not be discovered until damage has been. Diseases requiring drug treatment either singly or in combination are common in Africa, ranging from infections, endocrine diseases such as diabetes and thyroid diseases, haemoglobinopthies, Epilepsy, leprosy, etc.

Misoprostol: This is common drug used to induce abortion and is cheap, readily available and is sold over the counter in most African countries. Girls who have unwanted resort to its use without recourse to doctor's prescription. In approximately 80% of cases it fails to induce abortion and the pregnancy continue to term [14]. Misoprostol even though of low teratogenicity it is known to cause facial nerve paralysis, with or without limb defects, probably due to vascular disruption of the subclavian artery and an ischemia in the embryonic brain stem [15–17].

6. Anti-epileptics

Drugs such as Phenytoin and sodium valproate are known to cause birth defects.

Phenytoin: This is known to cause fetal hydantoin syndrome; IUGR, Microcephaly, Limb defects, Hypoplastic nails and distal phalanges, Heart defects and cleft lip [18]. Up to 1 in 10 of babies whose mothers take sodium valproate are at risk of having a birth defect and up to 1 in 40 have developmental problems as they grow. Sodium valproate use in pregnancy can cause; Spina bifida or a cleft palate, Atrial septal defect, Hypospadias, polydactyl and craniosynostosis [19].

Hydroxyurea: This drug is used in the management of Sickle cell disease and in acute myeloid leukemia. Sickle cell disease is common disease in Africa and has diverstating consequences and Hydroxyurea is used reduce crises and decrease anemia. This drug can cause fetal malformations including; Partial ossification of the cranial bones, absence of eye sockets, hydrocephalus, bipartite sternebrae and missing lumbar vertebrae.

Warfarin: Warfarin is used for anticoagulation and the fetuses of pregnant women on treatment with the drug may develop Fetal Warfarin syndrome, Hypoplasia of nasal bridge, Laryngomalacia, Pectus carinatus, Atrial septal defect, Patent Ductus arteriosus, Ventriculomegally, stippled epiphyses, telebrachydactyly and IUGR [20].

Alcohol: Alcohol consumption is very common in African communities and both local and bottled brews are consumed. This considered as traditional. Alcohol consumption in pregnancy is associated Fetal alcohol syndrome [FAS] characterized by; intellectual disability, behavior problems, IUGR and congenital heart defect can occur in an individual whose mother drunk alcohol during pregnancy [21]. I western Cape province of South Africa, more than 4% of 6- to 7-year-old school children had FAS. Comparable studies in Johannesburg found **2.7%** of children with FAS [22]. This raises concern about prevalence of FAS in middle and low-income countries where alcohol is available and used by women of reproductive age [23].

Traditional medicine consumption: Consumption of traditional medicine is common in Africa. These medicines are prepared from different herbs and other substances whose chemical composition is unknown. These preparations may contain chemical agents with teratogenic effects and cause congenital birth defects. Some birth defect which may be considered idiopathic may perhaps be caused by these traditional medications. Research is needed to determine the chemical constituents of some traditional medications to determine their teratogenic potential.

7. Deficiencies of essential elements and vitamins

Malnutrition as stated earlier is an important problem in African communities. Most of the essential elements are needed for normal fetal development. Malnourished women of reproductive age may lack these essential substances and could be at risk having babies with various forms of birth defects. An important vitamin whose deficiency leads to birth defect is folic acid and has been implicated as a cause of neural tube defect. Its use as a supplement in the preconception period and early pregnancy has been shown protect development of neural tube defects. Iodine deficiency is also implicated as a cause of birth defects [Iodine Deficiency Disorder characterized by intellectual disability motor and auditory disabilities] The severity depends on the level of deficiency in the mother. In 1998, an estimated 60,000 babies were born worldwide with severe iodine deficiency disorder (Cretinism), and an estimated 28 million pregnancies were still at risk of less severe Iodine-deficiency disorder from maternal iodine deficiency [24].

Exposure to pesticides/Herbicides: In Africa subsistence farming is source of livelihood and many women of reproductive age are engaged in it. Today there is a plethora of chemical agents used as pesticides and herbicides on the farm without wearing protective gear. As women are of the farming populations they are exposed to these potentially teratogenic chemicals and may thus be at risk of having their fetuses affected resulting in birth defects. There is the need for authorities to look in to the importance and use of these agents by instituting strict controls and regulations to catastrophic effects on the population.

Metabolic disease: Metabolic diseases such as diabetes have for been known to be associated poor pregnancy outcomes. Diabetic women are at risk of recurrent miscarriages, unexplained intrauterine fetal death, intrauterine growth restriction, fetal macrosomia and congenital birth defects. IDDM affects 0.5% of pregnancies in industrialized countries [25], and is becoming more prevalent in Africa especially among the elites and opulent segment of the population who are increasingly adopting western life styles. Infants born of mothers with insulin-dependent diabetes [IDDM] have up to threefold risk of having a serious birth defect [4]. Birth defect prevalence in infants of diabetic mothers is related to the level of control of the disease and in middle and low income countries diabetic control is sub-optimal because of low quality health infrastructure. Common birth defects seen infants whose mothers were diabetics include; Heart defects, Spinal and brain defects, Renal, Gastrointestinal tract defects and limb deficiencies.

How culture, beliefs and other factors influence prevalence of birth defects: Low-income countries particularly those in Africa are in special a position regarding the prevalence of certain diseases and conditions. In Africa women get pregnant at the extremes of age. Pregnancy at the age of 35 years and above is associated with an increased birth prevalence of chromosomal trisomies, particularly Down syndrome [4]. In middle and low-income countries, a high percentage of women give birth over the age of 35 years without the availability of community education and universally available and accessible family planning services, medical genetic screening, prenatal diagnosis, or associated services [4]. The birth prevalence of chromosomal aneuploidies is therefore high in these countries [26]. The percentage of births in women over 35 years ranges from 11 to 15% in developing regions of the world, compared to 5–9% in industrialized countries [27]. Another important factor affecting birth defect prevalence in Africa is the practice of consanguineous marriage. Consanguineous marriage is accepted by 20% of world's population [4]. This culture increases the birth prevalence of autosomal recessive birth defects, almost doubling the risk of neonatal and childhood death from birth defects [26, 28, 29]. Poverty is an important element in disease causation. Poor people are exposed to various deprivations leading to poor nutrition and susceptibility to infections and their attendant impact on general wellbeing. Mothers in poverty are more likely to be malnourished before and during pregnancy, and are at greater risk of exposure to environmental teratogens such as alcohol and maternal infections [30]. This will lead to congenital anomalies in the fetuses carried by those affected. In Africa malaria is endemic. Healthy carriers of recessive hemoglobin disorders (Sickle cell anemia and Thalassemia) glucose-6-phosphate dehydrogenase (G6PD) deficiency have a well-documented survival advantage against lethal effects of malaria compared to non-carriers of these conditions. As a result of this, carriers are more likely to survive to reproductive age. Over the years this has led to an increase in the population prevalence of these in tropical Africa [4]. Consequently, the birth prevalence of thalassemia, sickle cell disease and G6PD deficiency is high in malaria endemic regions of the world such as Sub-Saharan Africa, Eastern Mediterranean and North Africa, South East Asia and Western Pacific [31, 32].

Prenatal diagnosis its benefits and its challenges in Africa: The benefits of prenatal diagnosis can be viewed from different angles.

1. It provides an opportunity to classify the anomalies into lethal [e.g. Anencephaly, Bilateral renal agenesis] and non-lethal [e.g. Cleft lip, Polydactyl]

2. Surgically correctable and non-surgically correctable

3. It helps to identify those defects that are amenable to intra-uterine surgery and those that can be managed postnatally.

4. It helps decide when best to deliver, how to deliver and where to deliver

5. It provides us with a window to counsel the parents on the nature of the defect, treatment options and prognosis and thus assist them make an informed choice.

6. Where management is not available at the facility where diagnosis made, appropriate and timely referral can be made.

Prenatal diagnosis is testing for disease or condition in a fetus before it is born [1]. The aim of prenatal diagnosis is to detect birth defects which can morphological, genetic or biochemical. It involves different processes and it can be broadly classified in two, invasive and noninvasive. The invasive test requires taking fetal tissue which could be blood [Cordocentesis], placental tissue [Chorionic villus sampling] and amniotic fluid [Amniocentesis]. The non-invasive tests involve the use of ultrasound to image the various structures of the fetus to identify the normal or the abnormal. Ultrasound thus detects morphological aberrations [e.g. gastroschisis, omphalocele, anencephaly], or serve as a means of getting access to fetal tissues for further testing [CVS, Fetal amnio and Cordocentesis]. With further advancement in scientific techniques non-invasive test can now be done on maternal blood [Harvesting fetal cells in maternal blood and subjecting them to genetic testing]. In Africa prenatal diagnosis and screening for congenital defects is at the stage of infancy as the personnel and facilities are few and in most places non-existent, where available accessibility and affordability becomes an issue (**Table 3**).

	Urban Area Rural Area			
Geographical area	**One examination**	**≥ 3 examinations**	**One examination**	**≥ 3 examinations**
North Africa	88%	53%	20%	5%
Sub-Saharan Africa	32%	14%	6%	1%
Southern Africa	68%	38%	18%	6%

Source: Matres Mundi International Africa.

Table 3.
Prenatal ultrasonography in Africa.

Geographical area	**North Africa**	**Sub-Saharan Africa**	**Southern Africa**
Ultrasonographic PD	<50%	<10%	<25%
Biochemical PD	<5%	—	<5%
Invasive PD	<2%	—	<2%
Financial arrangement	Private	—	Private

Source: Foulkese Set al. [10].

Table 4.
Percentage of prenatal diagnostic techniques.

Prenatal diagnosis has positively impacted on our knowledge of congenital anomalies that is why it is essential even in low-income countries. Prenatal ultrasound diagnosis of congenital defects is one of the black points of African ultrasonography [33]. Ultrasound as an instrument for prenatal diagnosis is now available in many African countries but its application in prenatal diagnosis still faces a lot of challenges. The number of congenital defects diagnosed is very low and nearly always late in pregnancy [33]. This does not provide opportunity for meaningful and timely interventions. As there are no established screening programs in Africa, diagnosis of congenital defects is opportunistic and happens by chance. Malformation detection rates do not exceed 20% [34–36] (**Table 4**).

8. Reasons for the challenges

1. Poorly trained/untrained service providers

2. Inappropriate/obsolete equipment

3. Lack of ultrasound facilities in rural Africa where the bulk of the population reside

4. Absence of dedicated screening programs for congenital anomalies

5. Absence of laboratories for genetic and chromosomal analysis

6. Accessibility and affordability are important issues even where the services are available

In general, prenatal diagnosis for congenital anomalies is opportunistic and most often it happens by chance. Because of wide spread poverty the largest proportion of pregnant women are excluded from the opportunity.

9. Management of prenatally diagnosed congenital anomalies in Africa

Management/Treatment of any clinical condition including congenital anomalies is hinged on accurate and reliable diagnosis. Accurate diagnosis requires well trained personnel and appropriate equipment. A comprehensive management will also require the services of different specialists [Obstetricians, Neonatologists, Pediatric surgeons, special care nurses trained in care of infants with congenital anomalies]. These are all hard to come by in Africa. People generally recognize gross physical anomalies, hidden anomalies are not appreciated before birth [e.g. Cardiac anomalies]. It is only when the child is born and start manifesting with clinical symptoms that the parents will appreciate the problem. When a child is born with gross anomalies such anomalies may be associated with some syndrome. When the gross anomaly is corrected the genetic syndrome problem will remain and will manifest itself. Parents will attribute the manifestations of the genetic syndrome to metaphysical causes. It is thus difficult to make them understand the real cause and the possible remedy. As facilities for genetic/chromosomal analysis are few and in most cases non-existent it becomes difficult to make comprehensive evaluation and diagnosis. This makes final decision on management extremely difficult for the physician in Africa. Such is the environment perinatologists practice in Africa.

In this circumstance management of congenital anomalies cannot comprehensive and will be provided in a scattered manner. Often times when an anomaly is diagnosed treatment is limited to pregnancy termination where the laws allow. In one hospital 65% of pregnant women request for pregnancy termination when an anomaly is found in their fetus. In those with distressing polyhydramnius intermittent aspiration of the amniotic fluid is done to relieve the distress. Women who present with obstructed labor and a dead congenitally malformed fetus with hydrocephalus, delivery can be effected by craniotomy. For those that present with ruptured uterus, laparotomy is done and further management will depend on the extent of the rent and the clinical state of the fetus. Few countries [e.g. Egypt and South Africa] have centers that offer prenatal screening, diagnosis, treatment and follow up services, however such centers are not within the reach of the poor who carry most of the burden of congenital anomaly.

10. Illustrative cases

The cases below illustrate how pregnant women with undiagnosed fetuses with congenital anomaly will labor at home and developed ruptured uterus and present for treatment. All these cases presented with ruptured uterus following various interventions at home.

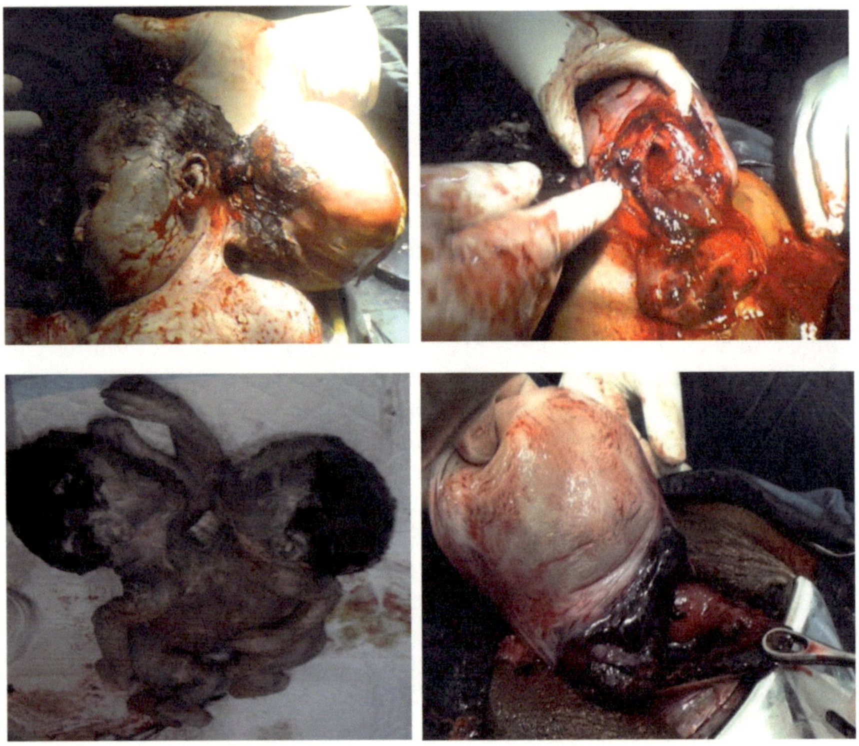

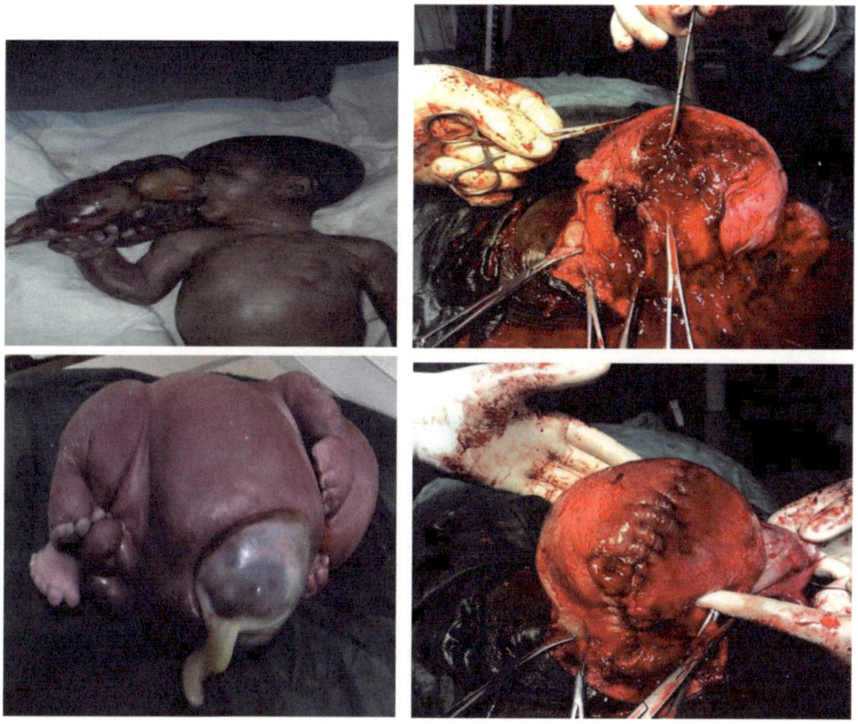

10.1 The way forward: collaborative care

This approach will require pooling of resources [Manpower, Equipment and other resources] to create referral hospitals in different countries and regions to serve as one stop shop able to provide care in all aspect of management of fetal congenital anomalies. Government, the private sector, philanthropists and other non-governmental organization can come together to establish such centers. Examples abound where similar collaboration has provided opportunity for treatment of some medical diseases requiring specialized care. The cardiothoracic center in Accra Ghana is now a regional center for referral that offer treatment for patients from different countries in West Africa as well as provide training for resident doctors from the whole sub-region. The first renal transplant at Aminu Kano Teaching Hospital was sponsored by philanthropist who also invites specialists from Britain and today thank to that effort the hospital is a referral and training center in renal transplant. In the management of congenital fetal anomalies similar approach can be adopted. The collaborative care group that I established at Abubakar Tafawa Balewa University Teaching Hospital Bauchi while I was there had achieved some success. We educated the community, Counsel parents, managed some pregnancies complicated by congenital anomalies and surgically treated a few cases despite the challenges we had. This can be replicated in other teaching hospitals.

11. Preventing birth defects: which approach?

The prevention of birth defects in Africa should be modified from the traditional approach that is adopted in other regions of the world. This is because of the peculiarities of the African environment. As much as possible all preventive strategies should be simple, low cost and innovative. All stakeholders must be involved, governments, communities, professional societies dealing with the issue, and non-governmental organizations. First stage should involve educating policy makers with emphasis on the burden of the problem on families and society at large. Educating policy makers on the need to understand how congenital anomalies contribute to both maternal and perinatal mortality. Areas that will require government intervention especially in community education, training of personnel and provision of equipment. Involving traditional and religious leaders who are gate keepers, educating them on the causes and prevention of birth defects. Involving the media in community enlightenment through discussion programs, talk shows and jingles.

Even as we want evolve our own model of prevention based on our peculiar social, environmental and economic circumstances, we must learn lessons from the experiences of other countries such as Cuba, China, India and Brazil. We must extract some of elements in the model they used and input them in to our own model. As we develop our preventive strategies we need to have the following at the background for us to succeed:

1. Endemic poverty and illiteracy

2. Uncontrolled birth rates and poor uptake of family planning services

3. Malnutrition and micronutrient deficiencies

4. Poor uptake of immunization services

5. Loose control of drugs and substances which are known to be teratogenic

6. Customs and traditions which are harmful to health

7. Our goal is to eliminate causative factors of congenital defect

8. Emphasis should be on primary prevention as we prepare to introduce secondary and tertiary prevention.

This is background to discussing the strategies for the prevention of congenital anomalies in Africa.

First strategy is education which is the backbone of development and progress in all spheres of life. It is known that the higher the level of literacy of a community the better are its economic indices, social status and health seeking behavior of its members. Education may be crucial in understanding the causes, treatment options and methods of prevention of congenital anomalies. It will also go along way in eliminating traditional and cultural practices that put communities at risk for congenital anomalies.

12. Specific measures

Immunization is an important strategy that has proven it efficacy in preventing other diseases. Congenital Rubella syndrome has virtually been eliminated in the United States and this is achieved through universal vaccination as a component of childhood immunization. In contrast the burden of congenital rubella in developing countries has been estimated to be about 100,000 per year [37], but only 28% of all developing countries have rubella immunization in place, as compared to 92% of industrialized countries [38]. Africa is the continent worse off compared to other regions as only one of 47 countries in Africa immunize against rubella. Currently there is an effort to develop a vaccine against Zika. For immunization program to succeed political will is required and thus policy makers should be encouraged by all to make commitment in this regard.

Folic acid and other micronutrients supplementation: Folate supplementation has been effective in preventing neural tube defect. Although available in various food substances in common use but one sure way of getting adequate levels in women before conception is through supplementation. Folate is chief and can be afforded by women in Africa. Different countries have used different approaches to ensure that women of reproductive age get adequate levels prior to conception. The United States Public Health Service recommended that women capable of becoming pregnant should consume 400 μg of folic acid daily [39], and in 1996 the Food and Drug Administration mandated the fortification of all enriched grain products, like flour and pastas, with 140 μg per 100 g of grain. The efficacy of folic acid in preventing neural tube defects has been proven in a community-based intervention study in China. In this study 400 μg of folic acid pills alone given before conception was found to be effective in reducing neural tube defects by 85% in an area of high prevalence and by 41% in an area of low incidence [40]. South American countries have implemented various successful folic acid supplementation programs [Chile and Cuba], supported by various organizations including the Pan American Health Organization, the March of Dimes and the CDC. There are plans being implemented to monitor the effect of the policy on the prevalence of neural tube defects, taking advantage of a preexisting and ongoing register of congenital malformations [41]. Africa can take queue to implement a similar program and follow up to determine its effectiveness in preventing neural tube defects in the African environment.

Family planning: Africa has one of the highest fertility rates in the world and this is creating concerns as it is overstretching resources in the continent. High fertility rates go hand in hand with prevalence of various pregnancy complications including birth prevalence of congenital malformation. To address this, Africa has to make effort to reduce its population growth. Uncontrolled population growth is a precursor to poverty, disease and malnutrition all of which contribute to the prevalence of congenital malformation. Family planning reduces the prevalence of congenital malformations by reducing birth rates and fertility. It has been estimated that in many countries with high fertility reducing the number of children per family to 2–3 could reduce the prevalence of genetic disorders by 40–50%. Further, combined with encouragement to complete reproduction before the age of 35 years, family planning can contribute to a 50% reduction of Down syndrome [42].

Avoidance of teratogens: Exposure to teratogens has multiple angles. In Africa use of traditional remedies is common in many communities. The chemical constituents of such remedies are not known and could contain teratogenic substances. Another factor is behavior of pharmaceutical companies selling drugs to consumers without following the regulatory procedures. People can also buy drugs over the counter without recourse to physician's prescription and use it anyhow with the potential of harm. Again counterfeit drugs are everywhere and people buy them because they are cheaper but their potential for harm is much greater compared to the original. Compounding these factors are lax environmental quality regulations and unhealthy working conditions which expose pregnant women to environmental pollutants [43]. Community education on what teratogens are, teratogens in the vicinity of communities and how they can cause harm can go along way in reducing exposure.

Other measures: Premarital counseling and testing can go along way in reducing genetic/hereditary disorders. A case in point is sickle cell disease in which premarital counseling and testing can reduce its prevalence. Avoiding consanguineous marriage has the potential of reducing propagation of hereditary diseases within population groups with hereditary diseases.

Secondary/tertiary prevention: This aims to reduce the number of children delivered with congenital malformation, whereas tertiary prevention is aimed at cure and amelioration of problems once a child with a congenital birth defect is born. Postnatal neonatal examination and screening of newborn children is a strategy in tertiary prevention, because once an anomaly is detected ameliorative measures can be instituted.

Ultrasound screening of congenital anomalies with an option of pregnancy termination: This has the potential of reducing the birth prevalence of congenital malformation but the issue of termination of pregnancy is the difficult part, as in many countries in Africa pregnancy termination can only be done when the life of the woman is at risk. Congenital malformations are not in themselves life threatening to the woman and therefore pregnancy cannot be terminated on account of congenital malformation in countries with restrictive abortion laws.

A new concept [Targeted screening]: The ideal thing is to screen all pregnant women for structural anomalies and test them for hereditary disorders and infections but the ideal is not always possible because the health care system in Africa is constrained by limited resource allocation. In view of this a transitional concept can be adopted pending acceptance of the whole population to prenatal screening and improve resource allocation to health care.

Targeted screening can be offered to the following category of pregnant women;

Women with; Previous history of babies with congenital anomalies, history of congenital anomaly in the family, index pregnancy with polyhydramnious, age more than 35 years, multiple gestation, consanguineous marriage, diabetes and those with sickle cell disease/Thalassemia.

13. Conclusion

Birth defects or congenital anomalies are important cause of perinatal mortality and morbidity. In developed countries successes were achieved in screening, treatment and prevention over the years. In Africa the picture is different as many factors play a role in causing congenital anomalies different from those seen in developed countries. In Africa factors such as poverty, illiteracy, malnutrition, exposure to teratogens and poor environmental control play an important role. Screening, treatment and preventive services for congenital anomalies are poorly developed. To achieve control primary prevention should be established and strengthened and when this is achieved, then secondary and tertiary control should follow. Innovative strategies should be employed in this endeavor.

Author details

Labaran Dayyabu Aliyu
Obstetrics and Gynecology Department, College of Medicine Bayero University, Kano, Nigeria

*Address all correspondence to: zainalabidinaliyu@yahoo.com

References

[1] World Health Organization [WHO], Congenital anomalies. 7 September 2016

[2] Akinyanju OO. A profile of sickle cell disease in Nigeria. Annals of the New York Academy of Sciences; 1989;**565**: 126-136

[3] Controlling Birth Defects: Reducing the Hidden Toll of Dying and Disabled Children in Low-Income Countries; Disease Control Priorities Project. December 2008

[4] Controlling birth defects: Reducing the hidden toll of dying and disabled children in low-income countries. Disease Control Priorities Project. The March of Dimes Global Report on Birth Defects 2006; 2018

[5] Congenital anomalies, Fact Sheet No. 370. Updated on January 2014

[6] Akinyanju OO, Otaigbe AI, Ibidapo MOO. Outcome of holistic care in Nigerian patients with sickle cell anaemia. Clinical Laboratory Hematology. 2005;**27**:195-199

[7] Congenital Malformations Worldwide. A Report from the International Clearinghouse for Birth Defects Monitoring Systems. Amsterdam: Elsevier; 1991

[8] Shamnas M, Arya PS, Thottumkal VA, Deepak MG. Congenital anomalies: A major public health issue in India. International Journal of Pharma and Bio Sciences. 2013;**3**(3):577-585

[9] The Millennium Development Goals Report. 2015

[10] National Demographic and Health Survey. Nigeria; 2018

[11] Turnpenny P, Ellard S. Emery's Elements of Medical Genetics. 12th ed. Edinburg, United Kingdom: Elsevier Churchill Livingstone; 2005

[12] Murry CJL, Lopez AD. Health Dimensions of sex and Reproduction: The Global Burden of Sexually Transmitted Diseases, HIV, Maternal conditions, Perinatal Disorders and Congenital Anomalies. Boston: Harvard School of Public Health; 1998

[13] WHO. Report of a Meeting on Preventing Congenital Rubella Syndrome: Immunization Strategies, Surveillance Needs. Geneva: WHO; 2000

[14] Journal of Community Genetics. Community Genet 2002;**5**:61-69

[15] Costa SH, Vessey MP. Misoprostol and illegal abortion in Rio de Janeiro, Brazil. Lancet. 1993;**341**:1258-1261

[16] Coelho HL, Teixeira AC, Santos AP, et al. Misoprostol and illegal abortion in Fortaleza, Brazil. Lancet. 1993;**341**: 1261-1263

[17] Pastuszak AL, Schuler L, Speck-Martins CE, Coelho KEFA, Cordello SM, Vargas F, et al. Use of misoprostol during pregnancy and Moebius' syndrome in infants. The New England Journal of Medicine. 1998;**338**:1881-1885

[18] Easton JD. Potential hazards of hydantoin use. Annals of Internal Medicine. 1972;**77**(6):988-989

[19] Jentink J, Loane MA, Dolk H, Barisic I, Garne E, Morris JK, et al. Workin EAS: Valproic acid monotherapy in pregnancy and major congenital malformations. The New England Journal of Medicine. 2010; **362**(23):2185-2193

[20] Hou J-W. Fetal warfarin syndrome. Chang Gung Medical Journal. 2004;**27** (9):3-5

[21] SAMHSA (Substance Abuse and Mental Health Services Administration).

"The Language of Fetal Alcohol Spectrum Disorders" Center for Substance Abuse Prevention. 2005. Available from: www.samhsa.org

[22] Croxford J, Viljoen D. Alcohol consumption by pregnant women in Western cape. South African Medical Journal. 1999;**89**:962-965

[23] Rosenthal J, Christianson AL, Cardero J. Fetal alcohol syndrome prevention in South Africa and other low-resource countries. American Journal of Public Health. 2005;**95**: 1099-1101

[24] UNICEF (United Nations International Children's Fund). The State of the World's Children. New York: UNICEF; 1998

[25] Khoury MJ, Becerra JE, Coredo JF, Erickson JD. Clinical-epidemiological assessment of pattern of birth defects associated with human teratogen: Application to diabetic embryopathy. Pediatrics. 1989;**84**:658-665

[26] WHO. Control of Hereditary Disease. WHO Technical Report Series 865. Geneva: WHO; 1996

[27] UNFPA. The State of the World Population. New York: United Nations Fund for population Activities; 1998

[28] Castilla EE, Gomez MA, Lopez-Camelo JS, Paz JE. The frequency of first cousin marriages on civil marriage certificates in Argentina. Human Biology. 1991;**63**:203-210

[29] Rittler MR, Liascovich JS, Lopez-Camelo, Castilla EE. Parental consanguinity in specific types of congenital anomalies. American Journal of Medical Genetics. 2001;**102**:36-43

[30] WHO/ICBDMS/EUROCAT. World Atlas of Birth Defects. 1st ed. Geneva: WHO; 1998

[31] Clegg JB, Weatheral DJ. Thalassemia and malaria: New insight into an old problem. Proceedings of the Association of American Physicians. 1999;**111**: 278-282

[32] Mokenhaupt FP, Ehrhardt S, Gellert S, Otchwemah RN, Dietz E, Anemana SD, et al. α+ thalassemia protects African children from severe malaria. Blood. 2004;**104**: 2003-2006

[33] Carrera JM. Obstetric ultrasound in Africa: Is it necessary to promote their appropriate use? Donald School Journal of Ultrasound in Obstetrics and Gynecology. 2011;**5**(3):289-296

[34] Adetiloye VA, Dare FO, Olayemi OA. A ten-year review encephalocele in a teaching hospital. International Journal of Gynecology & Obstetrics. 1993;**41**(3):241-249

[35] Adeleye AO, Dairo MD, Olowookere KG. Central nervous system congenital malformation in a developing country: Issues and challenges against their prevention. Childs Nervous System. 2010;**26**(7):914-924

[36] Maurali M, Fkih C, Essoussi-Chikhaoui J, Ben Haj Hassine A, Binous N, Ben Zineb N, et al. Gestational trophoblastic disease in Tunisia. La Tunisie Médicale. 2008; **86**(7):665-669

[37] Cutts FT, Wynnyck E. Modelling the incidence of congenital rubella syndrome in developing countries. International Journal of Epidemiology. 1997;**28**:1176-1184

[38] Robertson SE, Samuel R, et al. Control of rubella and congenital rubella syndrome (CRS) in developing countries. Part 2: Vaccination against rubella. World Health Organization. 1997;**75**:55-68

[39] CDC. Recommendations for the use of folic acid to reduce the number of

cases of spina bifida and other neural tube defects. MMWR. 1992;**41**(RR-14): 1-7

[40] Berry RJ, Zhu I, Erickson JD, Song I, et al. Prevention neural tube defects with folic acid in China. The New England Journal of Medicine. 1999;**341**: 1485-1490

[41] Nazer JH, Lopez-Camelo J, Castilla EE. ECLAMC: Results of thirty years of epidemiological surveillance of neural tube defects. Revista Médica de Chile. 2001;**129**:531-539

[42] WHO. Primary Health Care Approaches for the Prevention and Control of Congenital and Genetic Disorders. Geneva: WHO/HGN/WG/ 001; 2000

[43] Schuler-Faccini L, Leite JCL, Vieira-Sanseverino MT, Mizunski-Peres R. Evaluation of potential teratogens in Brazilian population. Ciência & Saúde Coletiva. 2002. In Press